Polytopes – Combinatorics and Computation

Gil Kalai
Günter M. Ziegler
Editors

Springer Basel AG

Editors:

Gil Kalai
Institute of Mathematics and Computer Science
The Hebrew University
Givat Ram
91904 Jerusalem
Israel

Günter M. Ziegler
Fachbereich Mathematik
MA 7-1
Technische Universität Berlin
Strasse des 17. Juni 135
D-10623 Berlin

2000 Mathematical Subject Classification 90C05, 65K05, 49M35, 52B05, 68Q25, 68U05, 65Y25

Library of Congress Cataloging-in-Publication Data
Polytopes ; combinatorics and computation / Gil Kalai, Günter M. Ziegler, editors.
      p. cm.   – (DMV Seminar ; Bd. 29)
    Includes bibliographical references.
    ISBN 978-3-7643-6351-2   ISBN 978-3-0348-8438-9 (eBook)
    DOI 10.1007/978-3-0348-8438-9
      1. Polytopes. I. Kalai, Gil, 1955- II. Ziegler, Günter M. III. Series.

  QA691 .P66 2000
  516.3'5--dc21

                                        00-034237

Deutsche Bibliothek Cataloging-in-Publication Data
Polytopes – combinatorics and computation / Gil Kalai ; Günter M. Ziegler. – Basel ; Boston ; Berlin :
Birkhäuser, 2000
   (DMV-Seminar ; Bd. 29)
   ISBN 978-3-7643-6351-2

© 2000 Springer Basel AG
Originally published by Birkhäuser Verlag in 2000

Printed on acid-free paper produced from chlorine-free pulp. TCF ∞
Cover design: Heinz Hiltbrunner, Basel

ISBN 978-3-7643-6351-2

9 8 7 6 5 4 3 2 1

DMV Seminar
Band 29

Contents

Preface

Questions that arose from Linear Programming and Combinatorial Optimization have been a driving force for modern Polytope Theory – cf. the diameter questions motivated by the desire to understand the complexity of the Simplex Algorithm, or the need to study facets for use in Cutting Plane procedures. But Algorithms are now driving Polytope Theory in a second way: we are now able to computationally study polytopes, to compute their parameters, and to construct examples of large complexity. It is still not always clear how to best utilize the data that are available from these new possibilities, but it is certainly clear that the new experimental tools open new doors and perspectives.

The papers of this volume display the rich connections that have thus changed Polytope Theory as a field: they highlight a meeting point of questions, ideas, results, algorithms and – not the least – computer programs from Discrete and Computational Geometry, Linear and Combinatorial Optimization, and Scientific Computing.

The volume grew from a DMV-Seminar "Polytopes and Optimization" in Oberwolfach, November 1997, and represents lectures and presentations from that workshop as well as additional invited papers. We are grateful to the Oberwolfach Research Institute for its hospitality, to the German-Israeli Foundation (GIF grant I-0309-146.06/93) for its support, and to everyone involved for their contributions and their enthusiasm.

Gil Kalai and Günter M. Ziegler
August 1999

Lectures on 0/1-Polytopes

Günter M. Ziegler*

Abstract. These lectures on the combinatorics and geometry of 0/1-polytopes are meant as an *introduction* and *invitation*. Rather than heading for an extensive survey on 0/1-polytopes I present some interesting aspects of these objects; all of them are related to some quite recent work and progress.

0/1-polytopes have a very simple definition and explicit descriptions; we can enumerate and analyze small examples explicitly in the computer (e. g. using `polymake`). However, any intuition that is derived from the analysis of examples in "low dimensions" will miss the true complexity of 0/1-polytopes. Thus, in the following we will study several aspects of the complexity of higher-dimensional 0/1-polytopes: the doubly-exponential number of combinatorial types, the number of facets which can be huge, and the coefficients of defining inequalities which sometimes turn out to be extremely large. Some of the effects and results will be backed by proofs in the course of these lectures; we will also be able to verify some of them on explicit examples, which are accessible as a `polymake` database.

Introduction

These lectures are trying to get you interested in 0/1-polytopes. But I must warn you: they are mostly "bad news lectures" – with two types of bad news:

1. General 0/1-polytopes are complicated objects, and some of them have various kinds of extremely bad properties such as "huge coefficients" and "many facets," which are bad news also with respect to applications.

2. Even worse, there are bad gaps in our understanding of 0/1-polytopes. Very basic problems and questions are open, some of them embarassingly easy to state, but hard to answer. So, 0/1-polytopes are interesting and remain *challenging*.

A good grasp on the structure of 0/1-polytopes is important for the "polyhedral combinatorics" approach of combinatorial optimization. This has motivated an extremely thorough study of some special classes of 0/1-polytopes such as the traveling salesman polytopes (see Grötschel & Padberg [31] and Applegate, Bixby,

* Supported by a DFG Gerhard-Hess-Forschungsförderungspreis (Zi 475/1-2) and by a German Israeli Foundation (G.I.F.) grant I-0309-146.06/93.

Cook & Chvátal [5]) and the cut polytopes (see Deza & Laurent [19], and Section 4). In such studies the question about properties of general 0/1-polytopes, and for complexity estimates about them, arises quite frequently and naturally. Thus Grötschel & Padberg [31] looked for upper bounds on the number of facets, and we can now considerably improve the estimates they had then (Section 2). One also asks for the sizes of the integers that appear as facet coefficients – and the fact that these coefficients may be huge (Section 5) is bad news since it means that there is a great danger of numerical instability or arithmetic overflow.

Surprisingly, however, properties of *general 0/1-polytopes* have not yet been a focus of research. I think they should be, and these lecture notes (expanded from my DMV-Seminar lectures in Oberwolfach, November 1997) are meant to provide support for this.

Of course, the distinction between "special" and "general" 0/1-polytopes is somewhat artificial. For example, Billera & Sarangarajan [9] have proved the surprising fact that *every* 0/1-polytope appears as a face of a TSP-polytope. Nevertheless, a study of the broad class of general 0/1-polytopes provides new points of view. Here it appears natural to look at *extremal* polytopes (e. g. polytopes with "many facets"), and at *random polytopes* and their properties.

Where is the difficulty in this study? The definition of 0/1-polytopes is very simple, examples are easy to come by, and they can be analyzed completely. But this simplicity is misleading: there are various effects that appear only in rather high dimensions ($d \gg 3$, whatever that means). Part of this we will trace to one basic linear algebra concept: determinants of 0/1-matrices, which show their typical behaviour – large values, and a low probability to vanish – only when the dimension gets quite large. Thus one rule of thumb will be justified again and again:

Low-dimensional intuition does not work!

Despite this (and to demonstrate this), our discussion in various lectures will take the low-dimensional situation as a starting point, and as a point of reference. (For example, the first lecture will start with a list of 3-dimensional 0/1-polytopes, which will turn out to be deceptively simple.)

However, examples are nevertheless important. The `polymake` project [28, 29] provides a framework and many fundamental tools for their detailed analysis. Thus, these lecture notes come with a library of interesting examples, provided as a separate section of the `polymake` database at

> http://www.math.tu-berlin.de/diskregeom/polymake/

We will refer to examples in this database throughout. The names of the polytope data files are of the form `NN:d-n.poly`, where `NN` is an identifyer of the polytope (e. g. initials of whoever supplied the example), `d` is the dimension of the polytope, `n` is its number of vertices. I invite you to play with these examples. (Also, I am happy to accept further contributions to extend this bestiary of interesting 0/1-polytopes!)

1. Classification of combinatorial types

1.1. Low-dimensional 0/1-polytopes

0/1-polytopes may be defined as the convex hulls of finite sets of 0/1-vectors, that is, as the convex hulls of subsets of the vertices of the regular cube $C_d = \{0,1\}^d$. Until further notice let's assume that we only consider full-dimensional 0/1-polytopes, so we have $P = P(V) = \text{conv}(V)$ for some $V \subseteq \{0,1\}^d$, where we assume that P has dimension d. We call two polytopes 0/1-*equivalent* if one can be transformed into the other by a symmetry of the 0/1-cube.

Now 0/1-polytopes of dimensions $d \leq 2$ are not interesting: we get a point, the interval $[0,1]$, a triangle, and a square.

The figure below represents the classification of 3-dimensional 0/1-polytopes $P \subseteq \mathbb{R}^3$ according to 0/1-equivalence. An arrow $P \longrightarrow P'$ between two of them denotes that P' is 0/1-equivalent to a *subpolytope* of P, that is, $P' \sim P(V')$ and $P = P(V)$ for some subset $V' \subseteq V$.

The full-dimensional 0/1-polytopes of dimension $d = 4$ were first enumerated by Alexx Below: There are 349 different 0/1-equivalence classes.

In dimension 5 there are exactly 1226525 different 0/1-equivalence classes of 5-dimensional 0/1-polytopes. This classification was done by Oswin Aichholzer [2]: a considerable achievement, which was possible only by systematic use of all the symmetry that is inherent in the problem.

In October 1998, Aichholzer completed also an enumeration and classification of the 6-dimensional 0/1-polytopes up to 12 vertices. The complete classification of all 6-dimensional 0/1-polytopes is not within reach: in fact, even the output, a non-redundant list of all combinatorial types would be so huge that it is impossible to store or search efficiently: and thus it would probably[1] be useless.

1.2. Combinatorial types

Many fundamental concepts of general polytope theory can be specialized to the situation of 0/1-polytopes. The following reviews the basic definitions and concepts. See for example [58, Lect. 0-3] for more detailed explanations.

0/1-polytopes, just as all other polytopes, can be described both in terms of their vertices ("$\mathcal{V}$-presentation") and in terms of equations and facet-defining inequalities ("$\mathcal{H}$-presentation"). However, for 0/1-polytopes the first point of view yields the name, it gives the natural definition, and thus it also determines our starting point.

---

[1] "Where a calculator like the ENIAC today is equipped with $18,000$ vacuum tubes and weighs 30 tons, computers in the future may have only $1,000$ vacuum tubes and perhaps weigh only $1\frac{1}{2}$ tons." – *Popular Mechanics*, March 1949, p. 258.

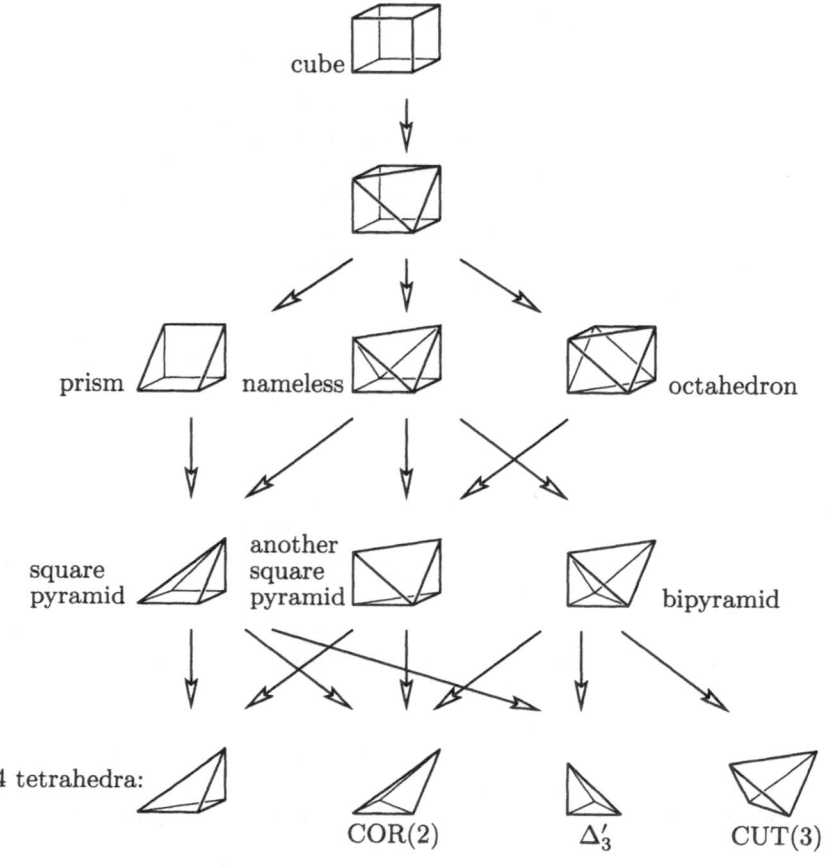

cube

prism   nameless   octahedron

square
pyramid   another
square
pyramid   bipyramid

4 tetrahedra:

COR(2) Δ'_3 CUT(3)

Definition 1 (0/1-polytopes). *A* 0/1-*polytope is a set* $P \subseteq \mathbb{R}^d$ *of the form*

$$P = P(V) \quad := \quad \text{conv}(V) \quad = \quad \{Vx : x \geq 0, 1^t x = 1\}$$

where $V \in \{0,1\}^{d \times n}$ *is a* 0/1-*matrix whose set of columns, a subset of the vertex set of the unit cube* $C_d = [0,1]^d$, *is the* vertex set *of* P.

Notation 2. Here and in the following, we will extensively rely on vector and matrix notation. Our basic objects are column vectors such as $x, y, \ldots$. Their transposed vectors x^t, y^t are thus row vectors. We use $\mathbf{1}$ to denote a column vector of all 1s (whose length is defined by the context), $\mathbf{0}$ to denote the corresponding zero vector, while e_i denotes the i-th unit vector (of unspecified length). The product $x^t y$ of a row with a column vector yields the standard scalar product, while xy^t is a product of a column vector with a row vector (of the same length), and thus represents a matrix of rank 1. Thus $\mathbf{1}^t \mathbf{1} = n$ if $\mathbf{1}$ has length n, while $\mathbf{11}^t$ is a square all-1s matrix. Matrices such as V and their sets of columns are used interchangeably. A unit matrix of size $n \times n$ will be denoted I_n.

It is hard to "see" what a 0/1-polytope looks like from looking at the matrix V. We have more of a chance to "understand" an example by feeding it to a computer and asking for an analysis. More specifically, we may present $P(V)$ to the polymake system of Gawrilow & Joswig [28, 29] in terms of a file that contains the key word POINTS in its first line, and then the matrix $(1, V^t)$ in the following lines – the rows of this matrix give homogeneous coordinates for the vertices of $P(V)$.

Example 3. For $n \geq 1$,

$$
\begin{aligned}
\Delta_{n-1} := P(I_n) &= \operatorname{conv}(\{e_1, \ldots, e_n\}) \\
&= \{x \in \mathbb{R}^n : x \geq 0, 1^t x = 1\} \subseteq \mathbb{R}^n
\end{aligned}
$$

is the *standard* simplex of dimension $n - 1$.
This is a *regular* simplex, since all its edges have the same length $\sqrt{2}$, but it is not full-dimensional, since it lies in the hyperplane given by $1^t x = 1$. Alternatively, we could consider the simplex

$$
\begin{aligned}
\Delta'_n = P(0, I_n) &= \operatorname{conv}(\{0, e_1, \ldots, e_n\}) \\
&= \{x \in \mathbb{R}^n : x \geq 0, 1^t x \leq 1\} \subseteq \mathbb{R}^n,
\end{aligned}
$$

which is full-dimensional, but not regular for $n \geq 2$. In fact, in many dimensions (starting at $n = 2$) there is no full-dimensional, regular 0/1-simplex at all. (See Problem 7.)

Example/Exercise 4. *For $V \in \{0, 1\}^{d \times n}$, let $\widetilde{V} = \binom{V}{I_n} \in \{0, 1\}^{(d+n) \times n}$. Then*

$$
P(\widetilde{V}) = \left\{ \binom{x}{y} \in [0, 1]^{d+n} : \quad y \geq 0, \ 1^t y = 1, \right.
$$

$$
\left. x_i = \sum_{j=1}^{n} v_{ij} y_j \ for \ 1 \leq i \leq d \right\}
$$

is an affine image of the $(n-1)$-dimensional standard simplex Δ_{n-1}. (Prove this!) Thus for the 0/1-polytope $P(\widetilde{V}) \subseteq \mathbb{R}^{d+n}$ we have a complete description in terms of linear equations and inequalities. From this we get $P(V)$ as the image of the projection

$$
\pi : \mathbb{R}^{d+n} \quad \longrightarrow \quad \mathbb{R}^d
$$

$$
\binom{x}{y} \quad \longmapsto \quad x
$$

that deletes the last n coordinates. Equivalently, to get $P(V) = \pi(P(\widetilde{V}))$ from $P(\widetilde{V})$ we must apply the operation "delete the last coordinate" n times.

Theorem 5 ($\mathcal{H}$-presentations). *Every 0/1-polytope $P(V) \subseteq \mathbb{R}^d$ can be written as the set of solutions of a system of linear inequalities, that is, as*

$$
P(V) = \{x \in \mathbb{R}^d : Ax \leq b\}
$$

for some $n \in \mathbb{N}$, a matrix $A \in \mathbb{Z}^{n \times d}$, and a vector $b \in \mathbb{Z}^n$.

Proof. First, we need not deal with equations in the system that describes $P(\tilde{V})$, since these can be rewritten in terms of inequalities: the equation $\boldsymbol{a}^t\boldsymbol{x} = \beta$ is equivalent to the two inequalities $\boldsymbol{a}^t\boldsymbol{x} \leq \beta$, $-\boldsymbol{a}^t\boldsymbol{x} \leq -\beta$. Thus, with the observations above, it suffices to show that if a set $S \subseteq \mathbb{R}^{k+1}$ has a description of the form

$$S = \left\{ \begin{pmatrix} \boldsymbol{x} \\ x_{k+1} \end{pmatrix} \in \mathbb{R}^{k+1} : \boldsymbol{a}_i^t\boldsymbol{x} + a_{i,k+1}x_{k+1} \leq b_i \ (1 \leq i \leq m) \right\},$$

then the projection of S to $\pi(S) \subseteq \mathbb{R}^k$ (by "deleting the last coordinate") has a representation of the same type. We may assume that the inequality system has been ordered so that

$$
\begin{aligned}
a_{i,k+1} > 0 \quad &\text{for} \quad 1 \leq i \leq i_0, \\
a_{j,k+1} < 0 \quad &\text{for} \quad i_0 < j \leq j_0, \\
a_{i,k+1} = 0 \quad &\text{for} \quad j_0 < i \leq m.
\end{aligned}
$$

Now for any given $\boldsymbol{x} \in \mathbb{R}^k$, it is easy to decide whether it lies in $\pi(S)$. Namely, $\boldsymbol{x} \in \pi(S)$ holds if and only if there is some value $\xi \in \mathbb{R}$ such that $\begin{pmatrix} \boldsymbol{x} \\ \xi \end{pmatrix} \in S$, where the inequalities for $1 \leq i \leq i_0$ provide upper bounds for such a value ξ, the inequalities for $i_0 < j \leq j_0$ give lower bounds, the others provide no conditions. Thus the system has a solution ξ for given $\boldsymbol{x}$ if all the upper bounds are at least as large as all the lower bounds. Explicitly, this yields a description of $\pi(S)$ as

$$
\left\{
\begin{aligned}
\boldsymbol{x} \in \mathbb{R}^k : (a_{i,k+1}\boldsymbol{a}_j - a_{j,k+1}\boldsymbol{a}_i)^t\boldsymbol{x} \leq a_{i,k+1}b_j - a_{j,k+1}b_i \quad & \text{for } 1 \leq i \leq i_0 \\
& \text{and } i_0 < j \leq j_0, \\
\boldsymbol{a}_i^t\boldsymbol{x} \leq b_i \quad\quad\quad\quad\quad\quad\quad\quad\quad\ \ & \text{for } j_0 < i \leq m
\end{aligned}
\right\},
$$

which is a presentation of the required form.   □

The transformation of an inequality system for S into a system for $\pi(S)$ in this way is known as *Fourier-Motzkin elimination* of the last variable [58, Lecture 1]. Note that, in the worst case, the system for $\pi(S)$ may have as many as $\left(\frac{m}{2}\right)^2$ inequalities: much more than the system for S! The good news at this point is that the inequality descriptions of $\pi(S)$ are typically very redundant: many of the inequalities can be deleted without changing the set of solutions of the system. However, the bad news is that even a minimal system – which in the case of a full-dimensional polytope P consists of exactly one inequality for each facet of P – may be huge. Correspondingly, 0/1-polytopes with rather few vertices may have "many" facets: See Section 2 below.

A projection argument together with the basic operation of "switching" will allow us for the following to assume that the polytopes under consideration are full-dimensional, and have $\boldsymbol{0}$ as a vertex, whenever that seems convenient:

(1) All the symmetries of the 0/1-cube $C_d = [0,1]^d$ transform 0/1-polytopes into 0/1-polytopes. In coordinates, these symmetries are generated by
- permuting coordinates, and

- replacing some coordinates x_i by $\bar{x}_i := 1 - x_i$ (*switching*).

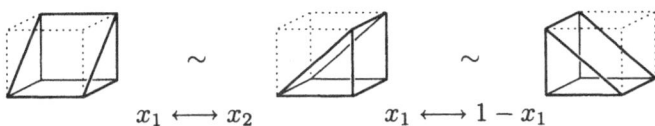

$$x_1 \longleftrightarrow x_2 \qquad\qquad x_1 \longleftrightarrow 1 - x_1$$

We call two 0/1-polytopes P and P' *0/1-equivalent* if a sequence of such operations can transform P into P'. In particular, one can transform any 0/1-polytope P with a vertex $v \in P \cap \{0, 1\}^n$ to a new, 0/1-equivalent polytope P' such that the vertex v gets mapped to the vertex $\mathbf{0}$ of P'.

(2) If $P \subseteq \mathbb{R}^{d+1}$ is not *full-dimensional*, then it is affinely equivalent to a 0/1-polytope $P' \subseteq \mathbb{R}^d$. To see this, first we may assume that $\mathbf{0} \in P$ (after switching), so P satisfies an equation of the form $a^t x + a_{d+1} x_{d+1} = 0$ with $a \in \mathbb{R}^d$.

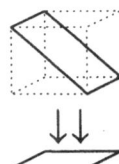

Furthermore, after permuting the coordinates we get that $a_{d+1} \neq 0$. But then "deleting the last coordinate"

$$\pi : \mathbb{R}^{d+1} \to \mathbb{R}^d$$

projects $P \to P' = \pi(P)$ injectively, that is, it defines an affine equivalence between P and $\pi(P) = P'$.

In the following, we usually deal with full-dimensional 0/1-polytopes, and we take 0/1-equivalence as the basic notion for their comparison. The resulting classification is much finer than the classification by affine equivalence – for example, all d-dimensional 0/1-simplices are affinely equivalent, but they are not necessarily 0/1-equivalent: Note that 0/1-equivalent polytopes are congruent, so they have the same edge lengths, volumes, etc. But the converse is not true, see below.

Definition 6. *The* faces *of a 0/1-polytope P are the subsets of the form $P^c = \{x \in P : c^t x = \gamma\}$, where $c^t x \leq \gamma$ is a linear inequality that is valid for all points of P. This definition of faces includes the subsets $\emptyset$ and P, the* trivial faces *of P.*

All faces of a 0/1-polytope are themselves 0/1-polytopes, of the form $F = \mathrm{conv}(F \cap \{0, 1\}^d)$. The set of 0-dimensional faces, or vertices, *of a 0/1-polytope is given by $V = P \cap \mathbb{Z}^d$. The 1-dimensional faces are called* edges. *Vertices and edges together form the* graph *of the polytope. The maximal non-trivial faces, of dimension $\dim(P) - 1$, are the* facets *of P. These are essential for the $\mathcal{H}$-presentation of polytopes: In the full-dimensional case every irredundant $\mathcal{H}$-presentation consists of exactly one inequality for each facet of P.*

The face lattice *is the set of all faces of P, partially ordered by inclusion. It is a graded lattice of length* $\dim(P)+1$. *Two polytopes are* combinatorially equivalent *if their face lattices are isomorphic as finite lattices.*

Proposition 7. *On the finite set of all 0/1-polytopes in $\mathbb{R}^d$ one has the following hierarchy of equivalence relations:*

$$\text{``0/1-equivalent''} \Rightarrow \text{``congruent''} \Rightarrow \begin{array}{c} \text{``affinely} \\ \text{equivalent''} \end{array} \Rightarrow \begin{array}{c} \text{``combinatorially} \\ \text{equivalent''} \end{array}.$$

For all three implications the converse is false, even when we restrict the discussion to full-dimensional polytopes.

Proof. The hierarchy is clearly valid: Every 0/1-equivalence is a congruence, congruent polyhedra are affinely equivalent, and affine equivalence implies combinatorial equivalence. In the following we provide counterexamples for all the converse implications.

(1) Full-dimensional 0/1-polytopes that are congruent but not 0/1-equivalent can be found in dimension 5:

VERTICES						VERTICES					
1	0	0	0	0	0	1	0	0	0	0	0
1	0	0	1	1	0	1	0	0	1	1	0
1	0	1	0	1	0	1	0	1	0	1	0
1	1	0	0	1	0	1	0	1	1	0	0
1	0	1	1	0	0	1	1	0	0	1	0
1	0	1	1	0	1	1	1	0	0	1	1

One easily checks that these two data sets (in `polymake` input format; see `CNG:5-6a.poly` and `CNG:5-6b.poly` in the `polymake` database) describe congruent, full-dimensional 0/1-simplices in $\mathbb{R}^5$: For this one just computes the pairwise distances of the points. A 0/1-equivalence would transform the array on the left to the array on the right by permuting rows and columns, and by complementing columns. But on the left we have two columns with exactly one 1 (and no column with five 1s), while on the right there is only one column with exactly one 1 (and no column with five 1s). Thus the two simplices are not 0/1-equivalent. Volker Kaibel has additionally shown that for $d \leq 4$ all congruent full-dimensional 0/1-polytopes are indeed 0/1-equivalent.

(2) The above classification for $d = 3$ contains examples of tetrahedra that are not congruent, but of course affinely equivalent. Further examples will appear in Lecture 2.

(3) Here are two 5-polytopes, `EQU:5-7a.poly` and `EQU:5-7b.poly`, that are combinatorially, but not affinely equivalent:

```
VERTICES                    VERTICES
1 0 0 0 0 0                 1 0 0 0 0 0
1 1 0 0 0 0                 1 1 1 0 0 0
1 0 1 0 0 0                 1 0 1 1 0 0
1 0 0 1 0 0                 1 0 0 1 1 0
1 0 0 0 1 0                 1 0 0 0 1 1
1 0 0 0 0 1                 1 1 0 0 0 1
1 1 1 1 1 1                 1 1 1 1 1 1
```

In fact, each of them is a bipyramid over a 4-simplex (and hence they are combinatorially equivalent), but in the first one the main diagonal is divided in the ratio $1 : 4$, for the other one the ratio is $2 : 3$, and such ratios are preserved by affine equivalences. □

1.3. Doubly exponentially many 0/1-polytopes

How many non-equivalent 0/1-polytopes are there? Clearly in $\mathbb{R}^d$ there are exactly 2^{2^d} different 0/1-polytopes, but some of them are low-dimensional, and some of them are equivalent to many others. Nevertheless, this trivial estimate is not that far from the truth.

For the following, let F_i^0 denote the facet of the d-cube $[0,1]^d$ that is given by $x_i = 0$, and similarly let F_i^1 be the facet given by $x_i = 1$. With a 3-dimensional picture in the back of our minds, we will refer to F_d^0 as the *bottom facet* and to F_d^1 as the *top facet* of C_d. All other facets will be called the *vertical facets* of C_d.

This terminology corresponds to one of the main proof techniques that we have for 0/1-polytopes: decomposition into "top" and "bottom" with induction over the dimension. For this we note the following for an arbitrary 0/1-polytope $P \subseteq [0,1]^d$:

- Every facet F_i^s induces a face $P_i^s := F_i^s \cap P$ of P; these faces are referred to as the *trivial faces* of P.
- Every vertex of P is contained either in the *bottom face* $P_d^0 = F_d^0 \cap P$ or in the *top face* $P_d^1 = F_d^1 \cap P$ of P.
- Every vertex v of P is determined by the set of trivial faces P_i^0 that contain it, since $v_i = 0$ holds if and only if $v \in P_i^0$.

The following figure (next page) illustrates that in general some trivial faces are facets, while others are not.

Proposition 8 (Sarangarajan-Ziegler). *There is a family $\mathcal{F}_d$ of $2^{2^{d-1}-4}$ different, full-dimensional 0/1-polytopes in $[0,1]^d$, such that*

- *any two polytopes in $\mathcal{F}_d$ are 0/1-equivalent if and only if they are combinatorially equivalent, and*
- *for $d \geq 6$, the collection $\mathcal{F}_d$ contains more than $2^{2^{d-2}}$ combinatorially non-equivalent d-dimensional 0/1-polytopes in $\mathbb{R}^d$.*

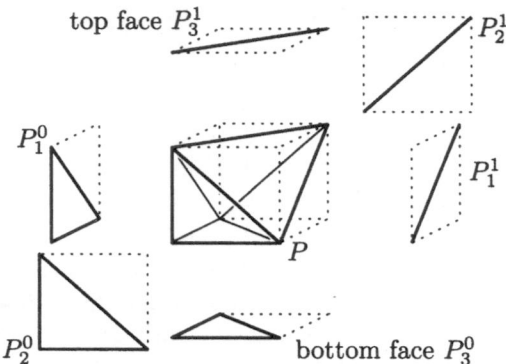

Proof. Let $d \geq 3$, and let $\mathcal{F}_d$ be the set of 0/1-polytopes $P(V) = \mathrm{conv}(V)$ of the following form:

- V contains all the vertices in the bottom facet F_d^0 of the d-cube $[0, 1]^d$ (that is, $\{0, 1\}^{d-1} \times \{0\} \subseteq V$),
- the pair $e_d, \mathbf{1}$ of opposite vertices of the top facet F_d^1 is contained in V,
- the two opposite vertices $e_d + e_1, \mathbf{1} - e_1$ of the top facet F_d^1 are *not* contained in V.

This fixes $2^{d-1} + 4$ vertices to be inside or outside V, and thus leaves $2^{2^d - (2^{d-1} + 4)} = 2^{2^{d-1} - 4}$ choices for the set V, and hence for the polytope $P(V)$.

For $d = 3$, there is exactly one polytope of the given special type (the "nameless" one):

Now the following facts are easy to verify about the polytopes $P(V) \in \mathcal{F}_d$:

- $P(V)$ is a d-dimensional 0/1-polytope. Its bottom facet $P_d^0 = F_d^0$ is a $(d-1)$-cube, with 2^{d-1} vertices.
- All the vertical facets F_i^s ($i < d$) induce facets P_i^s of $P(V)$. These are the facets of $P(V)$ that are adjacent to the cube facet P_d^0. Every vertex of $P(V)$ that is not on F_d^0 is completely determined by the set of vertical facets P_i^s that it lies on.
- All facets of $P(V)$, other than the bottom facet, have fewer than 2^{d-1} vertices. (For this we use that only the $2d + 2\binom{d}{2} = d^2 + d$ "special" hyperplanes given by $x_i = 0$, $x_i = 1$ or $x_i = x_j$ or $x_i = 1 - x_j$ contain 2^{d-1} 0/1-points, and all other hyperplanes contain less than 2^{d-1} 0/1-points. It is easy to verify that no special hyperplane other than "$x_d = 0$" can describe a facet of $P(V)$.)

- Therefore, if two polytopes $P(V)$ and $P(V')$ are combinatorially isomorphic, then they are equivalent by a symmetry of the d-dimensional 0/1-cube that fixes the bottom facet, and induces an automorphism of that bottom facet.
- The order of the symmetry group of C_{d-1} is $2^{d-1}(d-1)!$. So for each $P(V)$ there are not more than $2^{d-1}(d-1)!$ polytopes $P(V')$ that are combinatorially equivalent to it.
- Therefore, there are more than $2^{2^{d-1}-4}/(2^{d-1}(d-1)!)$ combinatorially non-isomorphic 0/1-polytopes of the form $P(V)$, and for $d > 5$ this number is larger than $2^{2^{d-2}}$.   □

2. The number of facets

2.1. Some examples

Staring too much at the 3-dimensional case, one might come up with the conjecture that a d-dimensional 0/1-polytope cannot have more than 2^d facets. In fact,

$$C_d^\Delta := \operatorname{conv}\{e_1, \ldots, e_d, 1 - e_1, \ldots, 1 - e_d\}$$

is a polytope with $2d$ vertices ($d \geq 3$) that is centrally symmetric with respect to $\frac{1}{2}\mathbf{1}$, the center of the 0/1-cube. Hence it is affinely equivalent to the usual regular d-dimensional cross polytope. In particular, this polytope has 2^d facets. The first examples are given as CRO:3-6.poly, CRO:4-8.poly, ... in the database.

(For $d = 4$ this construction produces a regular cross polytope CRO:4-8.poly, all of whose edges have length $\sqrt{2}$. Another remarkable regular cross polytope HAM:8-16.poly arises from the extended Hamming code $\widetilde{H}_8$. The cross polytopes C_d^Δ as constructed above are not regular for $d \neq 4$: they have edges of lengths $\sqrt{2}$ and $\sqrt{d-2}$.)

But more than that? Ewgenij Gawrilow was the first to detect a 5-dimensional 0/1-polytope with 40 facets. After intensive search, here is what we know about examples of low-dimensional 0/1-polytopes with "many facets" – and thus about #f(d), the maximal number $f_{d-1}(P)$ of facets that a d-dimensional 0/1-polytope P can have:

d	#f(d)	proved/found by	example
3	$= 8$		CRO:3-6.poly
4	$= 16$	Below	CRO:4-8.poly
5	$= 40$	Aichholzer	EG:5-10.poly
6	≥ 121	Sarangarajan	AS:6-18.poly
7	≥ 432	Christof	TC:7-30.poly
8	≥ 1675	Christof	TC:8-38.poly
9	≥ 6875	Christof	TC:9-48.poly
10	≥ 41591	Christof	TC:10-83.poly
$\vdots$	$\vdots$		$\vdots$
13	≥ 17464356	Christof	TC:13-254.poly

In brief: 0/1-polytopes may have *many* facets. But how many, at most? And how do 0/1-polytopes with "many facets" look like?

2.2. Some upper bounds

The asymptotically best upper bound for the number of facets of a d-dimensional 0/1-polytope is the following. I assume that it is rather tight; the problem is with the lower bounds, which look much worse.

Theorem 9 (Fleiner, Kaibel & Rote [26]). *For all large enough d, a d-dimensional 0/1-polytope has no more than*

$$\#f(d) \ \leq \ 30\,(d-2)!$$

facets.

See [26] for the (beautiful) proof of this result, which is probably valid for *all* d. The first bound of this order of magnitude was pointed out by Imre Bárány [58, p. 26]. Here we present a proof for the inequality

$$(*) \qquad\qquad \#f(d) \ \leq \ 2\,(d-1)! \ + \ 2(d-1),$$

which is asymptotically a bit worse than the one just quoted, but it is better in low dimensions – and whose proof (also from [26]) is strikingly simple.

For this, let $P \subseteq [0,1]^d$ be a d-dimensional 0/1-polytope. We note the following facts:

- The volume $\mathrm{Vol}_d(P)$ is an integral multiple of $\frac{1}{d!}$.
  (Every polytope can be triangulated without new vertices. Thus we are reduced to the case of 0/1-simplices, whose volume is given as $\frac{1}{d!}$ times the determinant – which is an integer.)

- The number of facets $f_{d-1}(P)$ of a d-dimensional 0/1-polytope P satisfies

$$f_{d-1}(P) \ \leq \ 2d \ + \ d!\,(1 - \mathrm{Vol}_d(P)).$$

  (This follows from an observation of Bárány: The d-cube $[0,1]^d$ has $2d$ facets. Now delete the "superfluous" 0/1-vectors, so that $[0,1]^d$ is gradually transformed into P. Whenever a facet of P "appears" in this process, a pyramid over the facet is removed, and the volume of this pyramid is at least $\frac{1}{d!}$.)

- Consider the projection $\pi : \mathbb{R}^d \longrightarrow \mathbb{R}^{d-1}$ that deletes the last coordinate. With respect to this projection, the boundary of P may be divided into "vertical," "upper" and "lower" facets. After projection, the images of the upper facets partition $\pi(P)$ into $(d-1)$-dimensional 0/1-polytopes, and so do the lower facets. Thus we get that

$$f_{d-1}^{\mathrm{upper}}(P),\ f_{d-1}^{\mathrm{lower}}(P) \ \leq \ (d-1)!\,\mathrm{Vol}_{d-1}(\pi(P)).$$

Our figure illustrates this decomposition for $d = 3$:

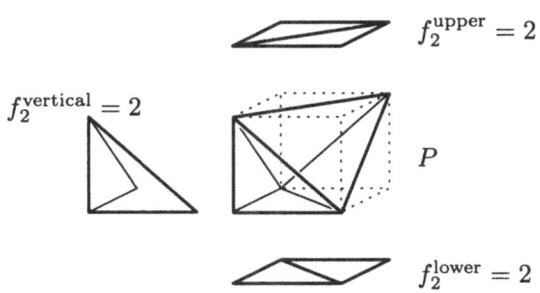

$f_2^{\text{upper}} = 2$

$f_2^{\text{vertical}} = 2$

P

$f_2^{\text{lower}} = 2$

- At the same time, the vertical facets of P are in bijection with a subset of the facets of $\pi(P)$: and the number of these can be estimated using the formula above:

$$f_{d-1}^{\text{vertical}}(P) \;\leq\; f_{d-2}(\pi(P)) \;\leq\; 2(d-1) + (d-1)!\,(1 - \underset{d-1}{\text{Vol}}(\pi(P))).$$

- ... and summing the upper bounds that we have obtained for $f_{d-1}^{\text{upper}}(P)$, $f_{d-1}^{\text{lower}}(P)$ and $f_{d-1}^{\text{vertical}}(P)$ completes the proof of $(*)$.   □

2.3. A bad construction

All the available data suggests that 0/1-polytopes may have much more than just simply-exponentially many facets. But no one has been able, up to now, to prove any lower bound on #f(d) that grows faster than c^d, for some constant $c > 1$.

Proposition 10 (Kortenkamp et al. [43]). *For all large enough d,*

$$\#f(d) \;>\; 3.6^d.$$

Proof. The *sum* $P_1 * P_2$ of two polytopes P_1 and P_2 is obtained by representing the polytopes in some $\mathbb{R}^n$ in such a way that their intersection consists of one single point which for both of them lies in the relative interior, and by then taking the convex hull:

$$P \;:=\; \text{conv}(P_1 \cup P_2), \quad \text{if} \quad P_1 \cap P_2 = \{x\} \quad \begin{array}{l}\text{is a relative interior point} \\ \text{for both } P_1 \text{ and } P_2.\end{array}$$

If we take the sum of two polytopes in this way, then the dimensions add, while the number of facets are multiplied. As an example, the sum of an n-gon (dimension 2, n facets) and an interval (dimension 1, 2 facets) results in a bipyramid over the n-gon (dimension 3, $2n$ facets). The sum operation is polar to taking products, where the dimensions add and the numbers of vertices are multiplied.

But we have to take a bit of care in order to adapt this general polytope operation to 0/1-polytopes, since there is very little space for "moving into a position" if we want to stay within the setting of 0/1-polytopes. For this call a 0/1-polytope *centered* if it has the center point $\frac{1}{2}\mathbf{1}$ in its (relative) interior. For example, among the 3-dimensional 0/1-polytopes, the 3-dimensional prism, the two different pyramids over a square, and the tetrahedra except for CUT(3), are *not* centered! On the other hand, the cross polytopes $C_d^\triangle$ are centered for all d.

The sum of two centered 0/1-polytopes $P_1 \subseteq [0, 1]^{d_1}$ and $P_2 \subseteq [0, 1]^{d_2}$ can be realized in $[0, 1]^{d_1+d_2}$ by embedding them into the subspaces $x_{d_1} = x_{d_1+1} = \ldots = x_{d_1+d_2}$ resp. $x_1 = \ldots = x_{d_1} = 1 - x_{d_1+1}$. This yields centered 0/1-polytopes $\widehat{P_1}, \widehat{P_2} \subseteq [0, 1]^{d_1+d_2}$ that are affinely isomorphic to P_1 and P_2, and whose convex hull realizes the sum $P_1 * P_2$. For example, the octahedron $C_3^\triangle$ can be viewed as the sum of a rectangle and a diagonal:

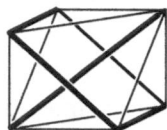

Now we need a starting block: and for that we use Christof's 13-dimensional 0/1-polytope TC:13-254.poly with at least $17464356 > 3.6^{13}$ facets. This polytope is indeed centered (you may check that already the first 22 vertices contain $\frac{1}{2}\mathbf{1}$ in their interior). Taking sums of copies of this polytope, and extra copies of $[0, 1]$ if needed, we arrive at the result. $\square$

This seems to be the best asymptotic lower bound available in the moment. I think that it is *bad*: one should be able to prove a lower bound of the form $c^{d \log d}$, or at least that there is a lower bound that grows faster than C^d for every $C > 1$! I'd offer two candidates for such a lower bound construction: Random polytopes, and cut polytopes. However, we cannot do the corresponding lower bound proof for either of these two classes, up to now.

Added in proof (March 2000): Imre Bárány and Attila Pór have recently achieved a lower bound of $\left(\frac{cd}{\log d}\right)^{d/4}$ for suitable random polytopes.

3. Random 0/1-polytopes

We do not understand random 0/1-polytopes very well. Let d be not too small, and take, say, $2d$ or $d \log d$ or d^2 random 0/1-points: *How will their convex hull look like? How many edges, and how many facets can we expect the random polytope to have?* We will see in this section that the analysis of random 0/1-polytopes is driven by one basic linear algebra parameter: the probability P_d that a random 0/1-matrix of size $d \times d$ has vanishing determinant.

This probability corresponds to the case of $d + 1$ random 0/1-points: Take $d + 1$ points $v_0, v_1, \ldots, v_d$ independently at random (where all 0/1 points appear with the same probability $p = \frac{1}{2^d}$). The $d+1$ points will be distinct with very high probability, and by symmetry we may assume that the first point is $v_0 = \mathbf{0}$. Thus the probability that the $d+1$ points span a d-dimensional simplex is exactly $1 - P_d$.

How large is this probability? We first study the case where d is small, and from this we will derive a quite misleading impression.

3.1. The determinant of a small random 0/1-matrix

Let P_d be the probability that a random 0/1-matrix of size $d \times d$ is singular. Of course we have

$$P_d = \frac{M_d}{2^{d^2}},$$

where M_d denotes the number of different 0/1-matrices of size $d \times d$ that have determinant 0. This number can be computed explicitly for $d \leq 7$:

d	M_d
1	1
2	10
3	338
4	42976
5	21040112
6	39882864736
7	292604283435872

(In fact, for $d \leq 6$ numbers that are equivalent to these were computed by Mark B. Wells in the sixties [44, p. 198]; the value for $d = 7$ is new, due to Gerald Stein.)

From this, we get a table for P_d, where for $d \geq 8$ we print estimates that were obtained by taking 10 million random matrices for each case:

$$
\begin{aligned}
P_1 &= 0.5 \\
P_2 &= 0.625 \\
P_3 &= 0.66015625 \\
P_4 &= 0.65576... \\
P_5 &= 0.62704... \\
P_6 &= 0.58037... \\
P_7 &= 0.51976... \\
P_8 &\approx 0.449 \\
P_9 &\approx 0.373 \\
P_{10} &\approx 0.298 \\
P_{11} &\approx 0.226 \\
P_{12} &\approx 0.164 \\
P_{13} &\approx 0.113 \\
P_{14} &\approx 0.075 \\
P_{15} &\approx 0.047
\end{aligned}
$$

Conclusion: the probability P_d first increases (!), but then it seems to decrease and approach 0 steadily, but not very fast.

3.2. Komlós' theorem and its consequences

The question about the probability P_d of singular random 0/1-matrices is equivalent to the same question about ± 1-matrices: P_d is equally the probability that a random ± 1-matrix of size $(d+1) \times (d+1)$ is singular. It is often convenient to switch to the ±1-case because it has more symmetry. The following proposition establishes the equivalence. Its observation is quite trivial, but also fundamental for various problems related to 0/1-polytopes.

Proposition 11 (Williamson [57]). *The map*

$$\varphi: A \longmapsto \begin{pmatrix} 1 & 1^t \\ 1 & 11^t - 2A \end{pmatrix} =: \widehat{A}.$$

establishes a bijection between the 0/1-matrices of size $d \times d$ and the ±1-matrices of size $(d+1) \times (d+1)$ for which all entries in the first row and column are $+1$.

The bijection φ satisfies $\det(\widehat{A}) = (-2)^d \det(A)$. In particular, it also provides a bijection between the invertible matrices of the two types.

Furthermore, there is a one-to-2^{2d+1} correspondence between the 0/1-matrices of size $d \times d$ and the ±1-matrices of size $(d+1) \times (d+1)$. The correspondence again respects invertibility.

Proof. Geometrically, the map φ realizes an embedding of $[0,1]^d$ as a facet of $[-1,+1]^{d+1}$. Algebraically,

$$\widehat{A} = \begin{pmatrix} 1 & 0^t \\ 1 & I_d \end{pmatrix}\begin{pmatrix} 1 & 1^t \\ 0 & -2A \end{pmatrix} \quad \text{arises from} \quad \begin{pmatrix} 1 & 1^t \\ 0 & -2A \end{pmatrix}$$

by adding the first row to all others, and thus we see that $\varphi(A)$ is indeed invertible if A is, and that $\det(\widehat{A}) = (-2)^d \det(A)$.

Finally, with every ±1-matrix one can associate a canonical matrix of the same size and type for which the first row and column are positive: for this first multiply columns by -1 in order to make the first row positive, then multiply rows to make the first column positive. There are exactly 2^{2d+1} matrices in $\{-1,+1\}^{(d+1)\times(d+1)}$ that have the same canonical form, corresponding to the $2d+1$ entries in the first row and column for which a sign can be chosen. □

Thus P_d measures for 0/1-matrices as well as for ±1-matrices the probability of determinant 0. Our experimental evidence is that P_d should converge to 0. But how fast? Here is what we know.

Theorem 12 (Komlós' Theorem; Kahn, Komlós and Szemerédi [40]).
The probability P_d that a random 0/1-matrix of size $d \times d$ is singular satisfies

$$\frac{d^2}{2^d} < P_d < 0.999^d$$

for all high enough d.

Proof. The non-trivial part is the upper bound, which is due to Kahn, Komlós and Szemerédi [40]. Their proof is difficult, involving a probabilistic construction. In fact, it is hard enough to prove that P_d converges to zero at all: this was first proved by Komlós in 1967 [42]; good starting points are Komlós' proof for $\lim_{d\to\infty} P_d = O(\frac{1}{\sqrt{d}})$ given in [12, Sect. XIV.2], and Odlyzko's paper [50].)

Here we only prove the lower bound. For this, we work in the ± 1-model, where P_d denotes the probability that a random $(d+1) \times (d+1)$-matrix is singular. In this model, the probability that two given rows are "equal or opposite" is $\frac{1}{2^d}$, and the same for two given columns. Altogether there are $2\binom{d+1}{2} = d^2 + d$ such events. These are not independent, but for any two such events the probability that they *both* occur is at most $\frac{1}{2^{2d-1}}$: if we look at two events that both refer to rows, or both refer to columns, then the probability that they both occur is $\left(\frac{1}{2^d}\right)^2$; if we want that two specific rows are equal or opposite, and two columns are equal or opposite at the same time, then the probability is $\left(\frac{1}{2^d}\right)\left(\frac{1}{2^{d-1}}\right)$. Thus we may estimate

$$P_d \;\geq\; (d^2 + d)\frac{1}{2^d} \;-\; \binom{d^2 + d}{2}\frac{1}{2^{2d-1}}$$

and this is larger than $\frac{d^2}{2^d}$ for $d > 10$. $\square$

It has been conjectured that the lower bound of this theorem is close to the truth:

Conjecture 13 (see [50], [40]). *The probability P_d that a random 0/1-matrix is zero is dominated by the possibility that one of the rows or columns is zero, or that two rows are equal, or two columns are equal:*

$$P_d \;\sim\; 2\binom{d+1}{2}\frac{1}{2^d} \;\sim\; \frac{d^2}{2^d}.$$

Equivalently: if a random ± 1-matrix of size $(d+1) \times (d+1)$ is singular, then "most probably" two rows or two columns are equal or opposite.

3.3. High-dimensional random 0/1-polytopes

Now we try to describe random 0/1-polytopes for large d.

Corollary 14. *With a probability that tends to 1 for $d \to \infty$ the following is true:*

   (i) *Any polynomial number of 0/1-vectors chosen (independently, with equal probability) from $\{0,1\}^d$ will be distinct.*
  (ii) *A set of d randomly chosen 0/1-points spans a hyperplane that does not contain the origin $\mathbf{0}$.*
 (iii) *The convex hull of $d+1$ random 0/1-points is a d-dimensional simplex.*

Proof. The probability for n random 0/1-vectors to be distinct is

$$\left(1 - \frac{1}{2^d}\right)\left(1 - \frac{2}{2^d}\right)\cdots\left(1 - \frac{n-1}{2^d}\right) \;>\; \left(1 - \frac{n}{2^d}\right)^n \;=\; \exp\left[n\ln\left(1 - \frac{n}{2^d}\right)\right],$$

and for $n \ll 2^d$ this can be estimated with $\ln(1 - \frac{n}{2^d}) \approx -\frac{n}{2^d}$, so we get a probability of at least $\exp(-\frac{n^2}{2^d})$, which converges to 1 if $\frac{n^2}{2^d}$ tends to zero.

If we choose $d+1$ random points, then by symmetry we may assume that the first one is $\mathbf{0}$. Thus the probability in question for the third statement, and also for the second one, is exactly $1 - P_d$, and thus both statements follow from Komlós' theorem. $\square$

But one would like to ask more questions. For example: *What is the expected volume of a random simplex?* It is indeed huge, as one can see from the following observations of Szekeres & Turán [56], see Exercise 7: for a random 0/1-matrix A of size $d \times d$, the expected value for the squared determinant is exactly[2]

$$E(\det(A)^2) \;=\; \frac{(d+1)!}{2^{2d}}.$$

But that means that 0/1-matrices A of determinant

$$|\det(A)| \;\geq\; \frac{\sqrt{(d+1)!}}{2^d}$$

exist, and are in fact common ("to be expected"). This is to be compared with the Hadamard upper bound

$$\det(\widehat{A}) \;\leq\; \sqrt{d+1}^{\,d+1}, \qquad \det(A) \;\leq\; \frac{\sqrt{d+1}^{\,d+1}}{2^d}.$$

that we will meet in Section 5.2.

Proposition 15 (Füredi [27]). *For any constant $\varepsilon > 0$, a random 0/1 polytope with $n \geq (2+\varepsilon)d$ vertices contains $\frac{1}{2}\mathbf{1}$, while a random polytope with $n \leq (2-\varepsilon)d$ vertices does not contain $\frac{1}{2}\mathbf{1}$, with probability tending to 1 for $d \to \infty$.*

Füredi's proof is elementary, combining Komlós' theorem with an estimate about the maximal number of regions in an arrangement of hyperplanes. Perhaps it can be adapted to prove that a random 0/1 polytope with $n \geq (2+\varepsilon)d$ vertices should even be centered?

Another question linked to Corollary 14 is: *Can we expect that there will be further 0/1-points on the hyperplane spanned by d random points?* We don't quite know, but the following result points towards an answer.

Proposition 16 (Odlyzko [50]). *With probability tending to 1 for $d \to \infty$, and*

$$n \;\leq\; d - \frac{10d}{\log d},$$

n random 0/1-points span an affine subspace of dimension n that does not contain any further 0/1-point.

One interesting question is whether this result could be extended to much bigger n. Of course, by Corollary 14(iii) to Komlós' theorem the statement fails (badly) if $n = d + 1$, but what about $n = d$? In other words, is there a high probability that d random 0/1-points will span a "simplex hyperplane"?

Still another, related question is: *If d random points span a hyperplane, is there a reasonable chance that this hyperplane is very unbalanced, with only few 0/1-points on one side?* This is closely linked (by "linearity of expectation") to the expected number of facets of a random polytope.

---

[2]The expected value for $\det(A)$ is 0 if $d > 0$, for symmetry reasons.

Proposition 17. *There is a constant $c > 0$ such that a random 0/1-polytope $P \subseteq$
$[0,1]^d$ with $n \leq (1+c)d$ vertices is "uniform" in the sense that any $d+1$ points
span a d-simplex, with probability tending to 1 for $d \to \infty$.*
(In particular, uniform polytopes are simplicial.)

Proof. Let $\gamma < 1$ be a constant such that $P_d \leq \gamma^d$ holds for all large enough d.
The probability that all $(d+1)$-subsets of a random sequence of n 0/1-vectors span
d-simplices is at least

$$\text{Prob}(P \text{ uniform}) \;\geq\; 1 - \binom{n}{d+1} P_d \;>\; 1 - \binom{(1+c)d}{d+1} \gamma^d$$

and with $(cd)! \approx \left(\frac{cd}{e}\right)^{cd}$ we estimate

$$\binom{(1+c)d}{d+1} \gamma^d \;\leq\; \frac{((1+c)d)^{cd}}{(cd-1)!}\gamma^d \;\approx\; \left(\frac{e(1+c)}{c}\right)^{cd}\gamma^d.$$

Thus $\text{Prob}(P \text{ uniform})$ will tend to 1 for large d if

$$\left(\frac{e(1+c)}{c}\right)^c \;<\; \frac{1}{\gamma}.$$

Thus by Theorem 12 one can take $c = 0.00009$. However, if Conjecture 13 were
true, then one could indeed take $c = 0.27$. $\qquad\square$

Note that if P is simplicial, then P_1^s is a simplex of dimension at most $d-1$
for $s = 0, 1$, and thus P has not more than $2d$ vertices. And simplicial polytopes
with $2d$ vertices do indeed exist: but the only examples that we know are centrally-
symmetric cross polytopes, which one gets as

$$\text{conv}\{v_1, v_2, \ldots, v_d, 1 - v_1, 1 - v_2, \ldots, 1 - v_d\},$$

where $v_1, v_2, \ldots, v_d \in \{0,1\}^d$ are d affinely independent points whose last coordinate is 0. Are there any other examples? This is not clear, but one may note that if
P is a d-dimensional cross polytope, then it must be centrally symmetric. In fact,
if v, w are vertices of P that are not adjacent, then they are not both contained
in any trivial face P_i^s (since these faces are simplices), hence they are opposite to
each other in the d-cube. But is every simplicial d-dimensional 0/1-polytope with
$2d$ vertices necessarily a cross polytope?

4. Cut polytopes

The "special" 0/1-polytopes studied in combinatorial optimization exhibit enormous complexity. One well-studied instance is that of the symmetric and asymmet-
ric travelling salesman (TSP) polytopes (see [31]), for which Billera and Sarangarajan [9] have shown that *all* 0/1-polytopes appear as faces.

In this lecture, we discuss basic properties of a different family of 0/1-poly-
topes, the cut polytopes, and of the correlation polytopes (a.k.a. boolean quadric

polytopes), which are affinely equivalent to them. For all of this and much more, Deza & Laurent [19] provides an excellent and comprehensive reference.

4.1. "Small" cut polytopes

Let's start with a "construction by example" of the "very small" cut polytopes; the general prescription will come in the next section.

A *cut* in a graph is any edge set of the form $E(S, V \setminus S) = E(V \setminus S, S)$, for $S \subseteq V$. That is, a cut consists of all edges that connect a node in S to a node not in S. For example, the complete graph K_3

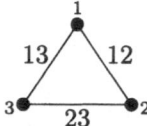

has four cuts: all the edge sets of size 2, as well as the empty set of edges:

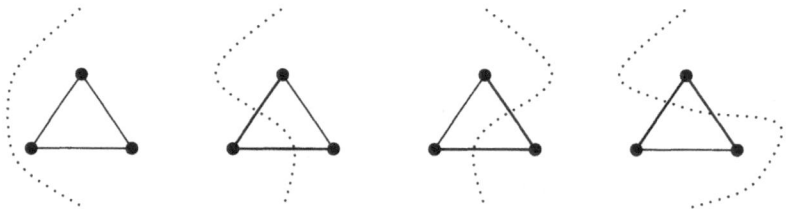

These cuts can be encoded by their *cut vectors*

$$\begin{pmatrix} x_{12} \\ x_{13} \\ x_{23} \end{pmatrix} \in \{0,1\}^3,$$

where the ij-coordinate records whether the edge ij is in the cut or not. The cut polytope is the convex hull of all these cut vectors. So, for K_3 we get the cut polytope `CUT3:3-4.poly` as

$$\mathrm{CUT}(3) = \mathrm{conv}\left\{ \begin{pmatrix} 0 \\ 0 \\ 0 \end{pmatrix} \begin{pmatrix} 0 \\ 1 \\ 1 \end{pmatrix} \begin{pmatrix} 1 \\ 0 \\ 1 \end{pmatrix} \begin{pmatrix} 1 \\ 1 \\ 0 \end{pmatrix} \right\}.$$

This 0/1-polytope is the convex hull of all 0/1-vectors of even weight (those just happen to be the cuts), so it is the regular simplex of side-length $\sqrt{2}$. Not a very interesting 0/1-polytope.

The complete graph K_4 has $\binom{4}{2} = 6$ edges, and altogether 8 cuts: the empty cut, the four cuts of size 3 that separate one vertex from the three others, and three cuts of size 4 that separate two vertices from the two others. Each cut yields a cut vector

$$(x_{12}, x_{13}, x_{14}, x_{23}, x_{24}, x_{34})^t \in \{0,1\}^6.$$

The resulting polytope `CUT4:6-8.poly` again has a very simple structure: it is a sum of two simplices,

$$\mathrm{CUT}(4) \;\cong\; \mathrm{CUT}(3) \;*\; \mathrm{CUT}(3) \;\cong\; \Delta_3 \;*\; \Delta_3.$$

To see this, note that four of the eight cuts of K_4

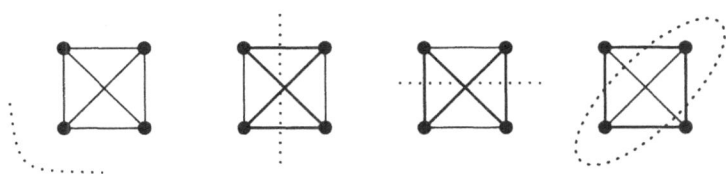

contain "none or both" from each pair of disjoint edges in K_4, that is,

$$x_{12} = x_{34}, \qquad x_{13} = x_{24}, \qquad x_{14} = x_{23},$$

so they lie in the 3-dimensional subspace U_1 of $[0,1]^6 \subseteq \mathbb{R}^6$ that is given by these three equations. The other four cuts (of size 3)

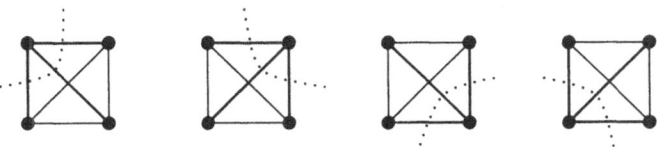

all contain exactly one edge from each disjoint pair, that is, they lie in the 3-dimensional subspace U_2 given by

$$x_{12} + x_{34} = 1, \qquad x_{13} + x_{24} = 1, \qquad x_{14} + x_{23} = 1$$

and give a 3-simplex that is equivalent to $\mathrm{CUT}(3)$ in this subspace. Now $U_1 \cap U_2 = \{\frac{1}{2}\mathbf{1}\}$ completes the analysis: we *understand the structure* of $\mathrm{CUT}(4)$. (Combinatorially, $\mathrm{CUT}(4)$ may also be identified with the cyclic polytope $C_6(8)$; in particular, it is simplicial, and neighborly. But nevertheless, it is not a very interesting polytope.)

And so on ... ? It turns out that the cut polytopes are much more complicated ("interesting") than one might think.

4.2. Cut polytopes and correlation polytopes

The definition/construction of the general cut polytopes follows a general method that has proved to be extremely successful in combinatorial optimization: The cuts in a complete graph K_n are encoded into the 0/1-polytope given by their characteristic vectors.

Definition 18 (Cut polytopes). *With every subset $S \subseteq [n] := \{1, \dots, n\}$, associate a 0/1-vector*

$$\delta(S) \in \{0,1\}^d, \quad d = \binom{n}{2},$$

by setting (for $1 \le i < j \le n$)

$$\delta(S)_{ij} := \begin{cases} 1 & \text{if } |S \cap \{i,j\}| = 1, \\ 0 & \text{otherwise.} \end{cases}$$

Thus we can identify the coordinates x_{ij} of $\mathbb{R}^d$ with the edge set of K_n (a complete graph with vertex set $[n]$), and the vector $\delta(S)$ represents the set $\{ij : x_{ij} = 1\}$ of edges ij of K_n that connect a vertex in S with a vertex in $\overline{S} := [n] \setminus S$, that is, a cut $E(S, \overline{S})$ in K_n.

The cut polytope CUT(n) *is defined by*

$$\text{CUT}(n) \quad := \quad \text{conv} \left\{ \delta(S) : S \subseteq [n] \right\} \quad \subseteq \mathbb{R}^d.$$

Lemma 19. *For every $n \ge 1$, and $d = \binom{n}{2}$, the cut polytope CUT(n) is a centered d-dimensional polytope with 2^{n-1} vertices.*

Proof. The two sets S and $\overline{S}$ determine the same cut $\delta(S) = \delta(\overline{S})$, but any two subsets $S, S' \subseteq [n-1]$ with $S \ne S'$ determine different cuts $\delta(S) \ne \delta(S')$, since $S = \{i \in [n-1] : \delta(S)_{in} = 1\}$. Thus

$$\text{CUT}(n) \quad = \quad \text{conv} \left\{ \delta(S) : S \subseteq [n-1] \right\} \quad \subseteq \mathbb{R}^d$$

has 2^{n-1} vertices (corresponding to the 2^{n-1} cuts of K_n). If CUT$(n) \subseteq \mathbb{R}^d$ were not full-dimensional, then it would satisfy some linear equation:

$$a^t x \;=\; \sum_{i,j} a_{ij} x_{ij} \;=\; \beta \quad \text{for all } x \in \text{CUT}(n)$$

for some non-zero $a \in \mathbb{R}^d$. However, the zero cut $\delta(\emptyset) = \mathbf{0} \in \text{CUT}(n)$ yields $\beta = 0$. Furthermore, we derive from the sketch below that

$$\delta(\{i\}) + \delta(\{j\}) - \delta(\{i,j\}) \;=\; 2e_{ij},$$

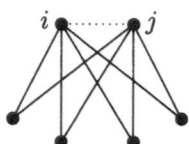

so $a^t \delta(S) = 0$ for all $S \subseteq [n]$ implies that

$$a^t(2e_{ij}) \;=\; 2a_{ij} \;=\; 0 \quad \text{for all } \{i,j\} \subseteq [n],$$

and thus $a = \mathbf{0}$.

To see that the cut polytopes are centered, it suffices to note any edge ij will be contained in a random cut with probability exactly $\frac{1}{2}$. Thus the average over all vertices of CUT(n) (that is, the centroid of the set of vertices) is $\frac{1}{2}\mathbf{1}$. $\square$

We note one more feature of the polytope CUT(n): it is very symmetric, with a vertex-transitive symmetry group. In fact, every symmetric difference of two cuts is a cut: this follows from the equation

$$E(S, \overline{S}) \vartriangle E(T, \overline{T}) \;=\; E(S \vartriangle T, \overline{S \vartriangle T}),$$

which is best verified and visualized in a little picture such as the following:

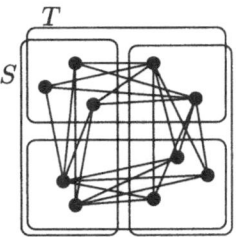

Thus for any $S \subseteq [n]$ the *switching map*

$$\sigma_S : \quad \mathbb{R}^d \to \mathbb{R}^d$$

$$x_{ij} \longmapsto \begin{cases} 1 - x_{ij} & \text{if } ij \in E(S, \overline{S}), \text{i. e., if } \delta(S)_{ij} = 1, \\ x_{ij} & \text{otherwise,} \end{cases}$$

defines an automorphism of $\mathrm{CUT}(n)$ that takes $\delta(T)$ to $\delta(T \triangle S)$, and thus takes the vertex $\delta(S)$ to the vertex $\delta(\emptyset) = \mathbf{0}$, and conversely. Thus, under such switching operations all vertices of $\mathrm{CUT}(n)$ are equivalent!

Next we will look at a different class of important 0/1-polytopes: the cut polytopes in (thin) disguise.

Definition 20 (Correlation polytopes). *The n-th correlation polytope is the convex hull of all $n \times n$ 0/1-matrices of rank 1:*

$$\mathrm{COR}(n) := \mathrm{conv}\{xx^t : x \in \{0,1\}^n\} \subseteq \mathbb{R}^{n^2}.$$

It is not so hard to see directly that $\mathrm{COR}(n)$ is a polytope of dimension $\binom{n+1}{2}$ with 2^n vertices, but the following observation yields even more.

Lemma 21 (de Simone [18]). *For $n \geq 2$ and $d = \binom{n}{2}$, there is a linear map*

$$\gamma : \mathbb{R}^{(n-1)^2} \longrightarrow \mathbb{R}^d$$

that induces an affine isomorphism of polytopes

$$\gamma : \mathrm{COR}(n-1) \cong \mathrm{CUT}(n).$$

Proof. For every correlation matrix xx^t we can extract the vector x from its diagonal, from this derive a set $S_x := \{i \in [n-1] : x_i = 1\}$, and thus get the cut vector $\delta(S_x)$. Furthermore, the components of $\delta(S_x)$ can be derived as *linear* combinations of the entries of xx^t:

$$\delta_{ij} := x_i(1 - x_j) + x_j(1 - x_i) = x_{ii} - x_{ij} + x_{jj} - x_{ji},$$

$$\text{and} \quad \delta_{in} := x_i = x_{ii}.$$

This defines a linear map $\gamma : \mathbb{R}^{(n-1)^2} \to \mathbb{R}^d$ which maps correlation matrices to cut vectors: $\gamma(xx^t) = \delta(S_x)$, and thus $\gamma(\mathrm{COR}(n-1)) = \mathrm{CUT}(n)$. An inverse map is obtained by taking

$$x_{ii} := \delta_{in}$$

$$\text{and} \quad x_{ij} = x_{ji} := \tfrac{1}{2}(x_{ii} + x_{jj} - \delta_{ij}) = \tfrac{1}{2}(\delta_{in} + \delta_{jn} - \delta_{ij}).$$

The image of this inverse map consists of only symmetric matrices in $\mathbb{R}^{(n-1)^2}$, which describes the $\binom{n}{2}$-dimensional subspace of $\mathbb{R}^{(n-1)^2}$ that is spanned by $\mathrm{COR}(n-1)$. □

Note that the isomorphism of Lemma 21 is *not* a 0/1-equivalence – in fact the polytopes are not 0/1-equivalent, even in their full-dimensional versions. For example cut polytopes are centered (Lemma 19), but the correlation polytopes are not: $\mathrm{COR}(n)$ contains the point $\frac{1}{2}\mathbf{1} = \frac{1}{2}(\mathbf{0}+\mathbf{1})$, but this point lies in the boundary, since $x_{11} \geq x_{12}$ is valid for all vertices of $\mathrm{COR}(n)$, and not for all of them with equality.

We now record a remarkable property of the correlation polytopes (and of cut polytopes, via Lemma 21):

Proposition 22. *Any three vertices of* $\mathrm{COR}(n)$ *determine a triangular face* $F \cong \Delta_2$, *that is,* $\mathrm{COR}(n)$ *is 3-neighborly.*

Proof. Using the symmetry of $\mathrm{CUT}(n+1)$, and its affine equivalence with $\mathrm{COR}(n)$, we may assume that one of the three vertices of $\mathrm{COR}(n)$ is $\mathbf{00}^t$, while the others are $\mathbf{uu}^t$ and $\mathbf{vv}^t$. The vectors $\mathbf{u}, \mathbf{v} \in \mathbb{R}^n$ span a 2-dimensional subspace $U \subseteq \mathbb{R}^n$, which may or may not contain a fourth 0/1-vector $\mathbf{y} \in \mathbb{R}^n$, but no fifth vector. However, if there is such a fourth vector $\mathbf{y}$, then we may assume that $\mathbf{y} = \mathbf{u} + \mathbf{v}$ (possibly after exchanging $\mathbf{y}$ with $\mathbf{u}$ or with $\mathbf{v}$).

Now take a *generic* vector $\mathbf{h} \in \mathbb{R}^n$ that is orthogonal to U – such a vector will satisfy $\mathbf{h}^t\mathbf{u} = \mathbf{h}^t\mathbf{v} = \mathbf{h}^t\mathbf{0} = 0$, and also $\mathbf{h}^t\mathbf{y} = 0$ if $\mathbf{y}$ exists, but $\mathbf{h}^t\mathbf{x} \neq 0$ for any other $\mathbf{x} \in \{0,1\}^n$. Then a little computation shows that the standard scalar product on $\mathbb{R}^{n^2}$ with $\mathbf{hh}^t$ defines a *linear* function on $\mathrm{COR}(n)$ that is minimized by $\mathbf{00}^t, \mathbf{uu}^t, \mathbf{vv}^t$, and by $\mathbf{yy}^t$ if this $\mathbf{y}$ exists, but by no other vertex of $\mathrm{COR}(n)$:

$$\langle \mathbf{hh}^t, \mathbf{xx}^t \rangle = \sum_{1 \leq i,j \leq n} (\mathbf{hh}^t)_{ij}(\mathbf{xx}^t)_{ij} = \sum_{1 \leq i \leq n} \sum_{1 \leq j \leq n} h_i h_j x_i x_j$$

$$= \Big(\sum_{1 \leq i \leq n} h_i x_i \Big)\Big(\sum_{1 \leq j \leq n} h_j x_j \Big) = (\mathbf{h}^t\mathbf{x})^2 \ \geq \ 0.$$

Now if there is no "fourth man" $\mathbf{y}$, then this proves that $\mathrm{conv}(\{\mathbf{uu}^t, \mathbf{vv}^t, \mathbf{00}^t\})$ is a (triangular) face of $\mathrm{COR}(n)$. If, however, $\mathbf{y} = \mathbf{u} + \mathbf{v}$ is present (that is, $\mathbf{u} + \mathbf{v} \in \{0,1\}^n$, and thus $\mathbf{u}^t\mathbf{v} = 0$), then we obtain that

$$F \ := \ \mathrm{conv}\,\big(\{\mathbf{00}^t, \mathbf{uu}^t, \mathbf{vv}^t, (\mathbf{u}+\mathbf{v})(\mathbf{u}+\mathbf{v})^t\}\big)$$

is a face of $\mathrm{COR}(n)$. We have to show that this face is a tetrahedron, not a 2-face.

Since $\mathbf{u}^t\mathbf{v} = 0$ with $\mathbf{u}, \mathbf{v} \neq \mathbf{0}$, we can take indices i, j with $u_i = 1$, $v_i = 0$ and $v_j = 1$, $u_j = 0$, so that

$$(\mathbf{uu}^t)_{ij} = u_i u_j = 0, \qquad (\mathbf{vv}^t)_{ij} = v_i v_j = 0, \qquad (\mathbf{yy}^t)_{ij} = (u_i + v_i)(u_j + v_j) = 1.$$

Thus $\mathbf{yy}^t$ cannot be linearly dependent of $\mathbf{uu}^t$ and $\mathbf{vv}^t$, and it is also clear that $\mathbf{uu}^t$ and $\mathbf{vv}^t$ are distinct 0/1-vectors and hence linearly independent. Thus $\mathbf{uu}^t, \mathbf{vv}^t$

and $(\boldsymbol{u} + \boldsymbol{v})(\boldsymbol{u} + \boldsymbol{v})^t$ are linearly independent, and hence F is a tetrahedron face of $\mathrm{COR}(n)$. □

This result is best possible, since $\mathrm{CUT}(n)$ is not 4-neighborly in general: for this we note (for $n \geq 3$) that

$$\delta(\emptyset) + \delta(\{1,2\}) + \delta(\{1,3\}) + \delta(\{2,3\}) = \delta(\{1\}) + \delta(\{2\}) + \delta(\{3\}) + \delta(\{1,2,3\})$$

which implies that the four vectors on either side of the equation (which are distinct vectors for $n \geq 4$) do *not* form a tetrahedron face of $\mathrm{CUT}(n)$.

Proposition 22 implies that $\mathrm{CUT}(n)$ is 5-*simplicial*, that is, all the 5-dimensional faces of $\mathrm{CUT}(n)$ are simplices (Exercise 7). On the other hand, the cut polytopes are not 6-simplicial: $\mathrm{CUT}(4)$ is 6-dimensional, but it is not a simplex. (Check SIMPLICIALITY for the cut polytopes in the polymake database!

Corollary 23. *For every dimension* $d = \binom{n}{2}$, *there is a 3-neighborly 0/1-polytope with more than* $2^{\sqrt{2d}-1/2}$ *vertices.*

Proof. Take $\mathrm{CUT}(n)$, whose number of vertices is 2^{n-1}, with $n = \frac{1}{2} + \sqrt{2d + \frac{1}{4}}$. □

4.3. Many facets?

Here I would also like to give – at least – a rough estimate of the number of facets of $\mathrm{CUT}(n)$ for large n, but that seems not that easy to get. We note that

$$\mathrm{CUT}(2) \cong \Delta_1 \quad \text{and} \quad \mathrm{CUT}(3) \cong \Delta_3$$

are simplices, while computation in "small" dimensions (see the polymake database and SMAPO [15]) yields

$\mathrm{CUT}(4)$ has dimension $d = 6$ and $16 = 1.5874^d$ facets,
$\mathrm{CUT}(5)$ has dimension $d = 10$ and $56 = 1.4956^d$ facets,
$\mathrm{CUT}(6)$ has dimension $d = 15$ and $368 = 1.4827^d$ facets,
$\mathrm{CUT}(7)$ has dimension $d = 21$ and $116764 = 1.7430^d$ facets,
$\mathrm{CUT}(8)$ has dimension $d = 28$ and $217093472 = 1.9849^d$ facets,
$\mathrm{CUT}(9)$ has dimension $d = 36$ and at least $12246651158320 = 2.3097^d$ facets.

This suggests that $\mathrm{CUT}(n)$ has more than d^{cd} facets, for some $c > 0$: prove this!

5. The size of coefficients

Grötschel, Lovász & Schrijver [30], in their study of the ellipsoid method and its (fundamental) role in optimization, introduced the notion of the *facet complexity* of a polyhedron. This is roughly the maximal number of bits that is necessary to represent one single facet by an inequality (with rational coefficients). They showed that for polyhedra with bounded facet complexity, optimization and separation are equivalent. Thus, the complexity of the facets is more important in this context than the number of facets. The following will imply that the facet complexity of an

n-dimensional 0/1-polytope is $O(n^2 \log n)$: this is a polynomial bound, and thus "good enough" for the ellipsoid method.

The question about the maximal facet complexity of 0/1-polytopes can also be phrased differently: it asks *How large integers (rationals) may occur in the $\mathcal{H}$-presentation of a 0/1-polytope?* The bad news is that the integer coefficients that appear in the inequality description of a 0/1-polytope may be *huge*. This is "bad": it means that all kinds of algorithms, from cutting plane procedures to convex hull algorithms – used to compute the facets of a given polytope – are threatened by "integer overflow" even in the case of 0/1-polytopes.

The main source for this lecture is a recent paper by Noga Alon and Văn H. Vũ [4], which rests on a construction of Johan Håstad [32] from 1992.

5.1. Experimental evidence

What do we mean by the size of the coefficients of the facets? For this we write each facet-defining inequality of a full-dimensional (!) 0/1-polytope uniquely in the normal form

$$\pm c_0 \pm c_1 x_1 \pm c_2 x_2 \pm \ldots \pm c_d x_d \;\geq\; 0,$$

for non-negative integers $c_0, c_1, \ldots, c_d$ with greatest common divisor 1. By the *greatest coefficient* we mean $\max\{c_1, \ldots, c_d\}$. For example, the inequality

$$19 - 12x_1 - 18x_2 - 3x_3 - 1x_4 + 10x_5 - 11x_6 + 4x_7 - 5x_8 \;\geq\; 0$$

from CF:8-9.poly has greatest coefficient 18.

The concept of greatest coefficient is invariant under permuting coordinates (obviously), but also under switching (the substitution $x_i \leftrightarrow 1 - x_i$ just switches the sign in front of $c_i x_i$, but not the size of the coefficient). It also changes the constant coefficient c_0, but we ignore these anyway. Note that for 0/1-polytopes we always have $c_0 \leq c_1 + \ldots + c_d$, since the facet-defining inequality must be satisfied by some 0/1-point with equality. We will, however, apply the concept of "greatest coefficients" only in the case of full-dimensional polytopes, since otherwise the "defining inequality of a facet" is not unique, which makes things more complicated.

With these precautions, we can look up the largest coefficient $\mathrm{coeff}(d)$ that appears in a facet-defining inequality for a d-dimensional 0/1-polytope, and for low dimensions d we find the following:

d	$\mathrm{coeff}(d)$	example
3	$= 1$	
4	$= 2$	CF:4-5.poly
5	$= 3$	CF:5-6.poly
6	$= 5$	CF:6-7.poly
7	$= 9$	CF:7-8.poly
8	$= 18$	CF:8-9.poly
9	≥ 42	CF:9-10.poly
10	≥ 96	CF:10-11.poly

Here the values for $d \leq 8$ are from complete enumeration, the values for $d > 8$ were taken from Aichholzer [1, p. 111]. The data for $d \leq 10$ do not, however, provide enough evidence to guess the truth.

5.2. The Alon-Vũ theorem and some applications

Let A be a 0/1-matrix of size $n \times n$. The question *How bad can A be?* has many aspects. Here we will first look (again) at the maximal size of a determinant $\det(A)$. Then we get to the Alon-Vũ theorem about the maximal size of entries of A^{-1}, and to its consequences for the arithmetics (large coefficients) and the geometry (e. g. flatness) of 0/1-polytopes.

Denote by ρ_n the maximal determinant of a 0/1-matrix of size $n \times n$. The exact value of ρ_n seems to be known for all $n < 18$, except for $n = 14$, where the following table quotes a conjecture.

n	ρ_n	
1	1	
2	1	
3	2	
4	3	
5	5	
6	9	
7	32	
8	56	
9	144	
10	320	
11	1458	
12	3645	
13	9477	
14	25515	(?, Smith [54], Cohn [16])
15	131072	
16	327680	...

Matrices that achieve these values may be obtained from a web page by Dowdeswell, Neubauer, Solomon & Turner [21].

Lemma 24 (The Hadamard bound). *The maximal determinant of a 0/1-matrix of size $n \times n$ is bounded by*

$$\rho_n \;\leq\; 2 \left(\frac{\sqrt{n+1}}{2} \right)^{n+1}$$

Proof. The Hadamard inequality states that the determinant of a square matrix is at most the product of the lengths of its columns, with equality (in the nonsingular case) if and only if all columns are orthogonal to each other. Applied to the case of a ± 1-matrix $\widehat{A}$ of size $(n+1) \times (n+1)$, this yields

$$(*) \qquad\qquad \det(\widehat{A}) \ \leq \ \sqrt{n+1}^{\,n+1}.$$

We transfer this result to $n \times n$ 0/1-matrices A via Proposition 11, and get

$$\det(A) \ \leq \ \frac{\sqrt{n+1}^{\,n+1}}{2^n},$$

as claimed.   □

A matrix $\widehat{A} \in \{-1, +1\}^{(n+1) \times (n+1)}$ that achieves equality in $(*)$ is known as a *Hadamard matrix*. It is not hard to show that for this a condition is that $n+1$ is $1, 2$ or a multiple of 4. It is conjectured that these conditions are also sufficient, but for many values $n+1 \geq 428$ this is not known. We refer to Hudelson, Klee & Larman [35] for an extensive, recent survey with pointers to the vast literature related to the Hadamard determinant problem. For the cases where $n+1$ is not a multiple of 4 one has slightly better estimates (by a constant factor) than the Hadamard bound; see Neubauer and Radcliffe [49]. Certainly for our purposes we may consider the Hadamard bound as "essentially sharp."

Now assume additionally that $A \in \{0, 1\}^{n \times n}$ is invertible (of determinant $\det(A) \neq 0$), consider the inverse $B := A^{-1}$, and let

$$\chi(A) := \max_{1 \leq i, j \leq n} |b_{ij}| = \|B\|_\infty$$

the largest absolute value of an entry of A^{-1}. These entries are – by Cramer's rule – given by

$$b_{ij} = (-1)^{i+j} \det(A_{ij}) / \det(A),$$

where A_{ij} is obtained from A by deleting the i-th row and the j-th column. Let $\chi(n)$ denote the maximal entry in the inverse of any invertible 0/1-matrix of size $n \times n$.

Theorem 25 (Alon & Vũ [4]). *The maximal absolute value of an entry in the inverse of an invertible 0/1-matrix of size $n \times n$ can be bounded by*

$$\frac{n^{n/2}}{2^{2n+o(n)}} \ \leq \ \chi(n) \ \leq \ \rho_{n-1} \ \leq \ \frac{n^{n/2}}{2^{n-1}}.$$

Furthermore, 0/1-matrices that realize the lower bound can be effectively constructed. (An even better lower bound, by a factor of 2^n, is achieved in the case where n is a power of 2.)

Before we look at the proof of this theorem, we derive two (quite immediate) applications to the geometry of 0/1-polytopes. First, let as above $\mathrm{coeff}(d)$ denote the largest c_i that can appear in a reduced inequality

$$\pm c_0 \pm c_1 x_1 \pm c_2 x_2 \pm \ldots \pm c_d x_d \ \geq \ 0,$$

that defines a facet of a d-dimensional 0/1-polytope in $\mathbb{R}^d$. (Here the c_i are non-negative integers, with $\gcd(c_1, \ldots, c_d) = 1$; by switching, we may assume that $c_0 = 0$ if we want to.)

Corollary 26 (Huge coefficients [4]). *The largest integer coefficient* $\mathrm{coeff}(d)$ *in the facet description of a full-dimensional 0/1-polytope in* $\mathbb{R}^d$ *satisfies*

$$\frac{(d-1)^{(d-1)/2}}{2^{2d+o(d)}} \;\leq\; \chi(d-1) \;\leq\; \mathrm{coeff}(d) \;\leq\; \rho_{d-1} \;\leq\; \frac{d^{d/2}}{2^{d-1}}.$$

Proof. Let $\{\mathbf{0}, \mathbf{v}_1, \dots, \mathbf{v}_{d-1}\} \subseteq \{0,1\}^d$ be points that span a hyperplane H in $\mathbb{R}^d$, and let $V = (\mathbf{v}_1, \dots, \mathbf{v}_{d-1})^t \in \{0,1\}^{(d-1)\times d}$. Then an equation that defines H is given by $\mathbf{c}^t\mathbf{x} = 0$, with $c_i = \pm\det(V_i)$, where $V_i \in \{0,1\}^{(d-1)\times(d-1)}$ is obtained from V by deleting the ith column. Thus we get the upper bound $\mathrm{coeff}(d) \leq \rho_{d-1}$ by definition.

For the lower bound $\chi(d-1) \leq \mathrm{coeff}(d)$ we start with a matrix $A \in \{0,1\}^{(d-1)\times(d-1)}$ such that $\chi(A) = |\det A_{11}/\det A| = \chi(d-1)$, and let $V := (A, \mathbf{e}_1) \in \{0,1\}^{(d-1)\times d}$. Then $|\det V_d| = |\det A|$, while $|\det V_1| = |\det A_{11}|$. Thus for the coefficients $c_i = \pm\det(V_i)$ of a corresponding inequality $\mathbf{c}^t\mathbf{x} \geq 0$ we get

$$|c_1/c_d| \;=\; |\det A_{11}/\det A| \;=\; \chi(d-1),$$

and thus for any integral inequality which defines a facet that lies in our hyperplane $H = \{\mathbf{x} \in \mathbb{R}^d : \mathbf{c}^t\mathbf{x} = 0\}$ we have $c_1 \geq \chi(d-1)$.

A simplex for which this H defines a facet is, for example, given by the convex hull of $\mathbf{0}$ and $\mathbf{e}_1$ together with the rows of V. This simplex has determinant $\det(A_{11})$, which will be huge for the matrices A constructed for the Alon-Vũ theorem. $\qquad\square$

Corollary 27 (Flat 0/1-simplices [4]). *The minimal positive distance* $\mathrm{flat}(d)$ *of a 0/1-vector from a hyperplane that is spanned by 0/1-vectors in* $\mathbb{R}^d$ *satisfies*

$$\frac{2^{d-1}}{\sqrt{d}^{\,d+1}} \;\leq\; \frac{1}{\sqrt{d}\,\rho_{d-1}} \;\leq\; \mathrm{flat}(d) \;\leq\; \frac{1}{\chi(d)} \;\leq\; \left(\frac{1}{d}\right)^{d/2} 2^{d(2+o(1))}.$$

Proof. Let $H = \mathrm{aff}\{\mathbf{0}, \mathbf{v}_2, \dots, \mathbf{v}_d\}$ be a hyperplane under consideration (we may assume that it contains the origin) and let $\mathbf{v}_1 \in \{0,1\}^d \setminus H$. Then there is an integral normal vector $\mathbf{c}$ to H with $c_i = \pm\det(A_{i1})$, for the square matrix $A := (\mathbf{v}_1, \mathbf{v}_2, \dots, \mathbf{v}_d) \in \{0,1\}^{d\times d}$. From $\mathbf{v}_1 \notin H$ we get $|\mathbf{v}_1^t\mathbf{c}| \geq 1$, while the length of $\mathbf{c}$ is bounded by

$$\|\mathbf{c}\| \;\leq\; \sqrt{d}\,\|\mathbf{c}\|_\infty \;\leq\; \sqrt{d}\,\rho_{d-1},$$

and thus

$$\mathrm{dist}(\mathbf{v}_1, H) \;=\; \frac{|\mathbf{v}_1^t\mathbf{c}|}{\|\mathbf{c}\|} \;\geq\; \frac{1}{\|\mathbf{c}\|} \;\geq\; \frac{1}{\sqrt{d}\,\rho_{d-1}}.$$

For the upper bound, take an A that achieves $\chi(A) = |\det(A_{11})/\det(A)| = \chi(d)$. Then

$$\frac{1}{\chi(d)} = \frac{|\det(A)|}{|\det(A_{11})|} = \frac{\mathrm{Vol}(\mathrm{conv}\{\mathbf{0}, \mathbf{v}_1, \mathbf{v}_2, \dots, \mathbf{v}_d\})}{\mathrm{Vol}(\mathrm{conv}\{\mathbf{0}, \mathbf{e}_1, \mathbf{v}_2, \dots, \mathbf{v}_d\})}$$

$$= \frac{\mathrm{dist}(\mathbf{v}_1, H)}{\mathrm{dist}(\mathbf{e}_1, H)} \geq \frac{\mathrm{dist}(\mathbf{v}_1, H)}{1},$$

where the last "=" is since we are considering two simplices with a common facet, and the inequality is from $\text{dist}(e_1, H) \leq \text{dist}(e_1, 0) = 1$. □

Proof. We now survey the main parts of the proof of the Alon-Vũ theorem, following [4].

(1) The upper bound. For the upper bound $\chi(n) \leq \rho_{n-1}$ we use that the entries of A^{-1} can be written as

$$b_{ij} = (-1)^{i+j} \frac{\det(A_{ij})}{\det(A)},$$

where the cofactors $A_{ij} \in \{0, 1\}^{(n-1) \times (n-1)}$ satisfy $|\det(A_{ij})| \leq \rho_{n-1}$ by definition, and the invertible matrix A satisfies $|\det(A)| \geq 1$ since it is integral.

(2) Super-multiplicativity. For the lower bound it is sufficient to construct "bad" matrices of size $2^m \times 2^m$, because of the following simple construction, which establishes

$$\chi(n_1 + n_2) \geq \chi(n_1) \cdot \chi(n_2).$$

Take "bad" invertible 0/1-matrices A and B of sizes $n_1 \times n_1$ and $n_2 \times n_2$, such that $\chi(A) = |\det A_{n_1,n_1} / \det A|$ and $\chi(B) = |\det B_{11} / \det B|$. Then the matrix

$$A \diamond B := \begin{pmatrix} A & & & & 0 & \\ & & & & & \\ 0 & \cdots & 1 & & & \\ \vdots & & \vdots & & B & \\ 0 & \cdots & 0 & & & \end{pmatrix}$$

has determinant $\det(A \diamond B) = \det(A) \cdot \det(B)$ and the submatrix

$$(A \diamond B)_{n_1, n_1+1} = \begin{pmatrix} & & & * & & \\ A_{n_1, n_1} & * & 0 & & \\ 0 & \cdots & 1 & * & * \\ \vdots & & \vdots & B_{11} & \\ 0 & \cdots & 0 & & \end{pmatrix}$$

has determinant $\det A_{n_1,n_1} \det B_{11}$, which establishes

$$\chi(A \diamond B) \geq \chi(A)\chi(B).$$

Thus – modulo an annoying computation that you may find in [4, Sect. 2.4] – it suffices to establish the lower bound of the Alon-Vũ theorem for $n = 2^m$.

(3) The construction. Here comes the key part of the proof: an ingenious construction of a "bad" ± 1-matrix whose size is a power of 2. Thus we prove that for $n = 2^m$ one can construct an invertible matrix $A \in \{+1, -1\}^{n \times n}$ with

$$\chi(A) = n^{n/2} \left(\frac{1}{2}\right)^{n+o(n)}$$

and then use Proposition 11. For this, the following is an explicit recipe. Perhaps you want to "do it" for $m = 3$, $n = 8$?

(i) Choose an ordering $\alpha_1, \alpha_2, \ldots, \alpha_n$ on the collection of all $2^m = n$ subsets of $[m] = \{1, 2, \ldots, m\}$, such that $|\alpha_i| \leq |\alpha_{i+1}|$ and $|\alpha_i \triangle \alpha_{i+1}| \leq 2$ holds for all i. This is not hard to do.

(ii) The matrix $Q \in \{+1, -1\}^{n \times n}$ given by $q_{ij} := (-1)^{|\alpha_i \cap \alpha_j|}$ is a symmetric Hadamard matrix (in fact, in lexicographic ordering of the rows and columns this is the "obvious" Hadamard matrix of order 2^m). Thus $Q^2 = nI_n$, $Q^{-1} = \frac{1}{n}Q$, and $\det(Q) = n^{n/2}$.

(iii) We construct a lower triangular matrix $L \in \mathbb{Q}^{n \times n}$ row-by-row, with $(1, 0, \ldots, 0)$ as the first row. For $i > 1$ define $A_i := \alpha_{i-1} \cup \alpha_i$ and

$$F_i := \begin{cases} \{\alpha_s : \alpha_s \subseteq A_i, \; |\alpha_s \cap (\alpha_{i-1} \triangle \alpha_i)| = 1 & \text{if } |\alpha_{i-1} \triangle \alpha_i| = 2, \\ \{\alpha_s : \alpha_s \subseteq A_i = \alpha_i\} & \text{if } |\alpha_{i-1} \triangle \alpha_i| = 1, \end{cases}$$

so that both $\alpha_{i-1}, \alpha_i \in F_i$ and $|F_i| = 2^k$ hold in both cases, for

$$k := |\alpha_i|.$$

Then for $1 < i \leq n$ and $1 \leq j \leq n$ we set

$$\ell_{ij} := \begin{cases} 0 & \text{if } \alpha_j \notin F_i, \\ \left(\frac{1}{2}\right)^{k-1} - 1 & \text{if } j = i - 1, \text{ and} \\ \left(\frac{1}{2}\right)^{k-1} & \text{otherwise.} \end{cases}$$

(iv) We define $A := LQ$. A simple computation shows that $a_{ij} \in \{+1, -1\}$ holds for all i, j. The determinant of A is 2^{n-1}, since $\det(Q) = n^{n/2} = 2^{m2^{m-1}}$ and

$$\det(L) = \prod_{i=1}^{n} \ell_{ii} = \prod_{k=1}^{m} \left(\frac{1}{2}\right)^{(k-1)\binom{m}{k}} = \left(\frac{1}{2}\right)^{\sum_{k=1}^{m}(k-1)\binom{m}{k}},$$

with

$$\sum_{k=1}^{m}(k-1)\binom{m}{k} = \sum_{k=1}^{m} k\binom{m}{k} - \sum_{k=1}^{m}\binom{m}{k} = m2^{m-1} - 2^m + 1.$$

Thus $|\det(A)|$ has the minimal possible value for an invertible 0/1-matrix of size $n \times n$.

(v) Take $i_0 := 2 + m + \binom{m}{2}$, which is the smallest index with $|\alpha_{i_0}| \geq 3$. We solve the system $Lx = e_{i_0}$. This is easy since L is lower triangular:
$x_i = 0$ for $i < i_0$,
$x_{i_0} = 1/\ell_{i_0 i_0} = 4$ since $\ell_{i_0 i_0} = \frac{1}{2^{3-1}} = \frac{1}{4}$,
and for $i > i_0$ we can solve recursively:

$$x_i = (2^{k-1} - 1)x_{i-1} - \sum_{\alpha_j \in F_i \setminus \{\alpha_i, \alpha_{i-1}\}} x_j \qquad \text{for } k = |\alpha_i|. \qquad (*)$$

Using induction, we now verify that the x_i are positive and

$$x_i > (2^{k-1} - 2)x_{i-1} \quad \text{for } i > i_0. \qquad (**)$$

Indeed, this holds for $i = i_0 + 1$, and by induction (with $k \geq 3$, so $2^{k-1} - 2 \geq 2$) we have

$$x_{i-1} > 2x_{i-2} > 4x_{i-3} > \ldots$$

Thus the sum in $(*)$ is smaller than

$$\frac{1}{2}x_{i-1} + \frac{1}{4}x_{i-1} + \ldots = \sum_{t \geq 1} \frac{1}{2^t} x_{i-1} = x_{i-1}.$$

Using this estimate in $(*)$ we get for $i > i_0$ that

$$x_i > (2^{k-1} - 1)x_{i-1} - x_{i-1} = (2^{k-1} - 2)x_{i-1}. \qquad (*\,*\,*)$$

Iteration of the recursion $(**)$, with a start at $x_{i_0} > 2$, now yields

$$x_n > \prod_{k=3}^{m} (2^{k-1} - 2)^{\binom{m}{k}} = \prod_{k=3}^{m} 2^{(k-1)\binom{m}{k}} \prod_{k=3}^{m} \left(1 - \frac{2}{2^{k-1}}\right)^{\binom{m}{k}}$$

where the first product is 2^N with

$$N = \sum_{k=1}^{m} (k-1)\binom{m}{k} - \binom{m}{2} = m2^{m-1} - 2^m - \binom{m}{2}$$

using the same sum as in (iv), and thus

$$2^N = 2^{m2^{m-1} - 2^m - \binom{m}{2}} = \frac{n^{n/2}}{2^{n + \binom{m}{2}}} = n^{n/2} \left(\frac{1}{2}\right)^{n+o(n)}.$$

Now we use that $1 - x \geq \frac{1}{2^{2x}}$ for $0 \leq x \leq \frac{1}{2}$ and thus estimate that the second product is at least $(\frac{1}{2})^M$ for

$$M = 2 \sum_{k=3}^{m} \frac{1}{2^{k-2}} \binom{m}{k} < 8 \sum_{k=0}^{m} \frac{1}{2^k} \binom{m}{k} = 8 \left(\frac{3}{2}\right)^m = 8\, n^{\log 3/2} = o(n).$$

Taken together, we have verified that

$$x_n = n^{n/2} \left(\frac{1}{2}\right)^{n+o(n)}.$$

(vi) The rest is easy: to get the i_0-th column of A^{-1}, we solve the system

$$Ay = e_{i_0} \iff LQy = e_{i_0} \iff Qy = x \text{ and } Lx = e_{i_0}.$$

But $Qy = x$ is easy to solve because of $Q^{-1} = \frac{1}{n}Q$. Thus we obtain

$$B_{ii_0} = y_i = \frac{1}{n} \sum_{j=1}^{n} q_{ij} x_j.$$

Here $|q_{ij}| = 1$ by construction and from $(***)$, for $k \geq 4$ $(n \geq 16)$, we have

$$x_n > 4x_{n-1} > 8x_{n-2} > \ldots$$

which yields

$$B_{ii_0} \;=\; y_i \;>\; \frac{1}{n}\left(\frac{1}{2}x_n\right) \;=\; \frac{1}{2n}x_n \;\geq\; n^{n/2}\left(\frac{1}{2}\right)^{n+o(n)}$$

Thus *all* entries of the i_0-column of A^{-1} are "huge." $\qquad\qquad\square$

5.3. More experimental evidence

The Alon-Vũ construction is completely explicit; you will find corresponding simplices (generated by Michael Joswig) as `MJ:16-17.poly` and as `MJ:32-33.poly` in the `polymake` database. The first one is a 16-dimensional simplex with "−451" appearing as a coefficient. The second one has dimension 32, and here you'll find tons of coefficients like "4964768222" that are indeed large enough to cause trouble for any conventional single-precision arithmetic system ...

6. Further topics

There are so many interesting aspects of 0/1-polytopes, and so little time and space. In this section, I am therefore collecting brief notes about three further topics, together with pointers to the literature that I'd hope you'll follow.

6.1. Graphs

General facts about graphs of polytopes apply in the 0/1-context, but there are new phenomena appearing – the most tantalizing perhaps being the Mihail-Vazirani conjecture. But we start with a basic fact that is true for all (bounded, convex) polytopes, and hence need not be proved in our more special context.

Theorem 28 (Balinski [6]; Holt & Klee [34]).

(1) *The graph of every d-dimensional polytope is vertex d-connected; that is, there are d vertex-disjoint paths between any pair of vertices.*

(2) *For any generic linear objective function (such that no two vertices get the same value), there are d monotone vertex-disjoint paths from minimum to maximum.*

In a setting of general (convex, bounded) polytopes the first part of this, "Balinski's Theorem," is a classic. The second part is a rather recent strengthening observed by Holt & Klee [34]: it implies the first part since for any two distinct vertices of a polytope we may assume that they are the unique minimal and the unique maximal vertex for a generic linear function, *after a projective transformation* [58, p. 74]. One peculiar phenomenon is that this reduction does not work in a setting of 0/1-polytopes: projective transformations do not preserve 0/1-polytopes.

The second result for this section is an example of an important and still unsolved problem from the theory of general polytopes (see [58, Sect. 3.3]) which becomes quite trivial when specialized to 0/1-polytopes – as was first noticed by Denis Naddef.

Theorem 29 (The Hirsch conjecture for 0/1-polytopes: Naddef [48]).
The diameter of the graph of a d-dimensional 0/1-polytope $P \subseteq \mathbb{R}^n$ is at most

$$\mathrm{diam}(G(P)) \leq d,$$

with equality if and only if P is (affinely equivalent to) a d-dimensional 0/1-cube. In particular, this implies that

$$\mathrm{diam}(G(P)) \leq n - d,$$

where n is the number of facets of P.

Proof. We get the first inequality by induction on dimension, the case $d = 1$ being trivial. If the two vertices in question lie in a common facet of $[0,1]^d$, then we can restrict to the corresponding trivial face of P of dimension at most $d - 1$, and we are thus done by induction. Hence we may assume that v and u are opposite vertices of $[0,1]^d$, and by symmetry only need to consider the case where $v = 0$ and $u = 1$.

But the vertex $u = 1$ is connected to some neighboring vertex u', and this neighbor is contained in some trivial face P_i^0, whose diameter is at most $d - 1$ by induction. Thus

$$d(v, u) \leq d(v, u') + d(u', u) \leq (d - 1) + 1 = d.$$

For the second statement, we may assume (using induction on dimension) that the two vertices in question do not lie on a common facet. Thus the polytope has at least $n \geq 2d$ distinct facets.   □

Our third item in this section is a conjecture that's plain wrong for general polytopes, but may be true in the 0/1-setting.

Conjecture 30 (Mihail-Vazirani [25, Sect. 7]). *The graph of every 0/1-polytope is a good expander. Specifically, for every partition $V = S \uplus \overline{S}$ of the vertex set, the polytope $P(V)$ has at least*

$$E(S, \overline{S}) \geq \min\{|S|, |\overline{S}|\}$$

edges between S and $\overline{S}$.

Remark: This may be very false. It does not seem to be trivial.

6.2. Triangulations

A very basic question is the following: *How many simplices are needed to triangulate the d-dimensional 0/1-cube?* Here the exact answer depends on the exact definitions: for example, let us assume that we want proper triangulations where all simplices are required to fit together face-to-face, and not only subdivisions, or (even worse) coverings. Let us also assume that we only admit triangulations without new vertices. (In general polytopes, new vertices *do* help – see Below et al. [8].)

In this setting, let triang(d) be the smallest number of simplices in a triangulation of $[0, 1]^d$. Then we can draw up a little table,

d	triang(d)
1	1
2	2
3	5
4	16
5	67
6	308
7	1493

combining many earlier results with those of Hughes [36, 37] and Hughes & Anderson [38, 39]. For $d = 8$, all we have seem to be the bounds $5522 \leq \text{triang}(8) \leq 11944$.

One of several curious effects in this context is that *not every* d-dimensional 0/1-polytope can be triangulated into at most triang(d) simplices: for example, for the 6-simensional half-cube `HC:7-64.poly`, the convex hull of all 0/1-vectors of even weight, one knows that the minimal number of simplices in a triangulation is $1756 > 1493 = \text{triang}(7)$ (Hughes & Anderson [39]).

A lower bound is certainly given by the maximal volume of a 0/1-simplex,

$$\text{triang}(d) \quad \geq \quad \frac{d!}{\rho_d},$$

but this bound is not very good. (For example, for $d = 3$ it yields only triang(d) $\geq$ 3). However, it can be refined by giving greater weight to the simplices "near the boundary," which have lower volume, but are needed to fill the 0/1-cube. A very elegant and powerful version of such a lower bound was given by Smith [55] using hyperbolic geometry.

A good quantity to consider is

$$\sqrt[d]{\frac{\#\text{simplices}}{d!}},$$

called the *efficiency* of a triangulation. This number is at most 1 for any triangulation that uses no "extra vertices." Haiman [33] showed that the limit

$$L \quad := \quad \lim_{d \to \infty} \sqrt[d]{\frac{\text{triang}(d)}{d!}}$$

exists, and that the efficiency of any example can also be achieved asymptotically, that is,

$$\sqrt[d]{\frac{\#\text{simplices}}{d!}} \quad \geq \quad L$$

holds for every triangulation without new vertices. The best upper bound on L up to now seems to be the one provided by Santos [52]:

$$L \leq \sqrt[3]{\frac{7}{12}} \approx 0.836.$$

One would, however, expect that the limit L is zero.

6.3. Chvátal-Gomory ranks

Interesting questions are related to the rounding procedures of integer programming that try to recover the convex hull $P_I := \text{conv}(P \cap \mathbb{Z}^d)$ from an inequality description of a polytope $P \subseteq [0,1]^d$.

In particular, Chvátal-Gomory rounding steps replace P by

$$P' := \bigcap_{H \supset P} H_I,$$

where the intersection is taken over all closed halfspaces H that contain P. The integer closure H_I of a halfspace H is easy to compute: make the left-hand side of the inequalities integral with greatest common divisor one, and then round the right-hand side. It was proved by Chvátal that a finite number of such closure operations lead from a bounded polytope P to its integer hull – *but how many steps are needed?* This quantity is known as the *Chvátal-Gomory rank* or *CG-rank* of the polytope P. We refer to the thorough treatment by Schrijver [53] for details and references.

Bockmayr, Eisenbrand, Hartmann & Schulz [11] noticed recently that for polytopes in the 0/1-cube, $P \subseteq [0,1]^d$ the Chvátal-Gomory rank is bounded by a polynomial in d. An improvement of Eisenbrand & Schulz [23] establishes that for $P \subseteq [0,1]^d$ the CG-rank is bounded by

$$(1+\varepsilon)d \leq \text{CGr}(d) \leq 3d^2 \log(d)$$

for some $\varepsilon > 0$.

But how about a good lower bound? Riedel [51] has implemented a procedure to compute the CG-rank for polytopes, and he has provided explicit, low-dimensional examples $P \subseteq [0,1]^d$ for which the CG-rank exceeds the dimension; so we know

$$\begin{aligned}
\text{CGr}(3) &= 3 \\
\text{CGr}(4) &= 5 \\
\text{CGr}(5) &= 6 \ (?) \\
\text{CGr}(6) &\geq 8 \\
\text{CGr}(7) &\geq 9
\end{aligned}$$

But can anyone provide a lower bound that is more than simply linear?

7. Problems and exercises

1. Is it true that every simple 0/1-polytope is a product of simplices? (This question was answered by Kaibel & Wolff [41].)

2. Estimate the maximal vertex degree of a d-dimensional 0/1-polytope. (Hint: `OA:5-18.poly`)

3. Classify the 0/1-polytopes of diameter $\sqrt{2}$.

4. *Bound the maximal number of vertices for a d-dimensional 2-neighborly 0/1-polytope. (Corollary 23 yields an exponential lower bound.)

5. Show that every 0/1-polytope without a triangle face is a d-cube. (Volker Kaibel noticed that this follows from a result of Blind & Blind [10].)

6. *Is it true that a simplicial 0/1-polytope of dimension d has at most 2^d facets? Is it true that every simplicial 0/1-polytope of dimension d with $2d$ vertices is centrally symmetric and thus is a cross polytope with exactly 2^d vertices? (This is true for $d \le 6$, according to Aichholzer's enumerations.)

7. Estimate the probability that the determinant of a random $(n \times n)$-matrix with entries in $\mathbb{Z}_2$ vanishes (for large n). Compare your result with that claimed in [46].

8. Show that for every fixed $\varepsilon > 0$, all the trivial faces of a random 0/1-polytope with $(2 - \varepsilon)d$ vertices are simplices, with probability tending to 1 for $d \to \infty$. (Volker Kaibel)

9. Prove the Szekeres-Turán theorem: The expected value of the determinant $\det(C)$ of a random ± 1-matrix $C \in \{-1, +1\}^{n \times n}$ is zero, but the expected value of the squared determinant is exactly $n!$:

$$E(\det(C)^2) \quad = \quad n!.$$

Hint, by Bernd Gärtner: Use $\det(C) = \sum_{i=1}^{n}(-1)^{i-1}c_{1i}\det(C_{1i})$, and analyze the expected values of the summands in

$$\det(C)^2 \quad = \quad \sum_{i=1}^{n}(\det(C_{1i}))^2 \; + \; \sum_{i \neq j}(-1)^{i+j}c_{1i}c_{1j}\det(C_{1i})\det(C_{1j}).$$

10. What is the largest absolute value of the determinant of an $n \times n$ matrix with coefficients in $\{-1, 0, 1\}$? With coefficients in the interval $[0, 1]$? With coefficients in $[-1, 1]$?
(It is reported that this is a question that was asked by L. Collatz at an international conference in 1961, and answered a year later by Ehlich & Zeller [22]. Your answers should be in terms of ρ_n resp. ρ_{n-1}.)

11. Show that $\mathrm{CUT}(k)$ is (0/1-isomorphic to) a face of $\mathrm{CUT}(n)$, for $k \le n$.

12. Prove that $[\mathbf{0}, \mathbf{1}]$ is an edge of the correlation polytope $\mathrm{COR}(n)$.

13. Show that $\mathrm{CUT}(n)$ is 6-*simplicial*: every 5-dimensional face is a simplex.

14. Show that the *metric polytope*

$$\text{MET}(n) := \left\{ \begin{array}{ll} X \in [0,1]^d : & x_{ij} - x_{ik} - x_{jk} \leq 0 \text{ and} \\ & x_{ij} + x_{ik} + x_{jk} \leq 2 \text{ for distinct } i, j, k \in [n] \end{array} \right\}$$

is an *LP*-relaxation of CUT(n): it satisfies CUT(n) = conv(MET(n) $\cap \mathbb{Z}^d$), where $d = \binom{n}{2}$ is the dimension of CUT(n) $\subseteq$ MET(n) $\subseteq \mathbb{R}^d$.
*Estimate the CG-rank of MET(n).

15. How do the inequalities for a 0/1-polytope transform into the inequalities for the corresponding $(+1/-1)$-polytope?

16. Give more and better examples of "large" coefficients appearing in the facet-defining inequalities of 0/1-polytopes.

17. Show that every triangulation of $\Delta_k \times \Delta_\ell$ without new vertices has exactly $\binom{k+\ell}{k}$ facets.

18. For which dimensions $d > 1$ and integers k $(1 \leq k \leq d)$ does there exist a regular d-dimensional 0/1-simplex of edge length $\sqrt{k}$?
    (Show that this is equivalent to the famous Hadamard determinant problem.)

19. *For which d is there a regular d-dimensional 0/1-cross polytope?

20. For which $E \subseteq \binom{[n]}{2}$ is $P(E) = \text{conv}\{e_i + e_j : \{i,j\} \in E\}$ a simplex? Show that every such simplex of dimension $d = \binom{n}{2}$ has normalized volume $\frac{2^k}{d!}$ for some $k \geq 0$. [17]

21. Estimate the volumes of the Birkhoff polytopes

$$B_{n+1} := \{X \in [0,1]^{n \times n} : \mathbf{1}^t X \leq \mathbf{1}^t, \ X\mathbf{1} \leq \mathbf{1}, \ \mathbf{1}^t X \mathbf{1} \geq n-1\}.$$

(see `BIR3:4-6.poly`, `BIR4:9-24.poly`, ...). The exact value of the volume of B_{n+1}, which is some integer divided by $n^2!$, is known for $n \leq 7$, due to Chan & Robbins [13].

Acknowledgements. Thanks to Bernd Gärtner, Carsten Jackisch, Fritz Eisenbrand, Francisco Santos, Gerald Stein, Imre Bárány, Jiří Matoušek, Marc Pfetsch, Michael Joswig, Noga Alon, Oswin Aichholzer, Thomas Voigt and Volker Kaibel for so many helpful discussions and useful comments.

References

[1] O. AICHHOLZER: *Hyperebenen in Huperkuben – Eine Klassifizierung und Quantifizierung*, Diplomarbeit am Institut für Grundlagen der Informationsverarbeitung, TU Graz, October 1992, 117 Seiten.

[2] O. AICHHOLZER: *Extremal properties of 0/1-polytopes of dimension 5*, in this volume, pp. 111–130

[3] O. AICHHOLZER & F. AURENHAMMER: *Classifying hyperplanes in hypercubes*, SIAM J. Discrete Math. **9** (1996), 225–232.

[4] N. ALON & V. H. VŨ: *Anti-Hadamard matrices, coin weighing, threshold gates, and indecomposable hypergraphs*, J. Combinatorial Theory, Ser. A **79** (1997), 133–160.

[5] D. Applegate, R. Bixby, V. Chvátal & W. Cook: *On the solution of traveling salesman problems*, in: "International Congress of Mathematics" (Berlin 1998), *Documenta Math.*, Extra Volume ICM 1998, Vol. III, 645–656.

[6] M. L. Balinski: *On the graph structure of convex polyhedra in n-space*, Pacific J. Math. **11** (1961), 431–434.

[7] F. Barahona & A. R. Mahjoub: *On the cut polytope*, Math. Programming **36** (1986), 157–173.

[8] A. Below, U. Brehm, J. A. De Loera & J. Richter-Gebert: *Minimal simplicial dissections and triangulations of convex 3-polytopes*, Preprint, ETH Zurich, August 1999, 15 pages.

[9] L. J. Billera & A. Sarangarajan: *All 0-1 polytopes are travelling salesman polytopes*, Combinatorica **16** (1996), 175–188.

[10] G. Blind & R. Blind: *Convex polytopes without triangular faces*, Israel J. Math. **71** (1990), 129–134.

[11] A. Bockmayr, F. Eisenbrand, M. Hartmann & A. S. Schulz: *On the Chvátal rank of polytopes in the 0/1-cube*, Preprint; *Discrete Applied Math.*, to appear.

[12] B. Bollobás: *Random Graphs*, Academic Press, New York 1985.

[13] C. Chan & D. P. Robbins: *On the volume of the polytope of doubly stochastic matrices*, Experimental Math., to appear.

[14] T. Christof: *SMAPO – Library of linear descriptions of small problem instances of polytopes in combinatorial optimization*, `http://www.iwr.uni-heidelberg.de/iwr/comopt/soft/SMAPO/SMAPO.html`

[15] V. Chvátal: *Edmonds polytopes and a hierarchy of combinatorial problems*, Discrete Math. **4** (1973), 305–337.

[16] J. H. E. Cohn: *On determinants with elements ±1, II*, Bulletin London Math. Soc. **21** (1989), 36–42.

[17] J. A. De Loera, B. Sturmfels & R. R. Thomas: *Gröbner bases and triangulations of the second hypersimplex*, Combinatorica **15** (1995), 409–424.

[18] C. De Simone: *The cut polytope and the boolean quadric polytope*, Discrete Math. **79** (1989/90), 71–75.

[19] M. M. Deza & M. Laurent: *Geometry of Cuts and Metrics*, Algorithms and Combinatorics **15**, Springer-Verlag, Berlin Heidelberg 1997.

[20] M. Deza, M. Laurent & S. Poljak: *The cut cone III: On the role of triangle facets*, Graphs and Combinatorics **8** (1992), 125–142; updated **9** (1993), 135–152.

[21] R. Dowdeswell, M. G. Neubauer, B. Solomon & K. Tumer: Binary matrices of maximal determinant, web page at `http://www.imrryr.org/~elric/matrix`.

[22] H. Ehlich & K. Zeller: *Binäre Matrizen*, Z. Angewandte Math. Physik **42** (1962), T20–T21.

[23] F. Eisenbrand & A. S. Schulz: *Bounds of the Chvátal rank of polytopes in the 0/1 cube*, in: Proc. IPCO '99, 137–150.

[24] G. Elekes: *A geometric inequality and the complexity of computing the volume*, Discrete Comput. Geometry **1** (1986), 289–292.

[25] T. FEDER & M. MIHAIL: *Balanced matroids*, in: Proceedings of the 24th Annual ACM "Symposium on the theory of Computing" (STOC), Victoria, Brithsh Columbia 1992, ACM Press, New York 1992, pp. 26–38.

[26] T. FLEINER, V. KAIBEL & G. ROTE: *Upper bounds on the maximal number of facets of 0/1-polytopes*, European J. Combinatorics **21** (2000), 121–130.

[27] Z. FÜREDI: *Random polytopes in the d-dimensional cube*, Discrete Comput. Geometry **1** (1986), 315–319.

[28] E. GAWRILOW & M. JOSWIG: *Polymake: a software package for analyzing convex polytopes*,
http://www.math.tu-berlin.de/diskregeom/polymake/doc/

[29] E. GAWRILOW & M. JOSWIG: *Polymake: a framework for analyzing convex polytopes*, in this volume, pp. 43–74

[30] M. GRÖTSCHEL, L. LOVÁSZ & A. SCHRIJVER: *Geometric Algorithms and Combinatorial Optimization*, Algorithms and Combinatorics **2**, Springer-Verlag, Berlin Heidelberg 1988.

[31] M. GRÖTSCHEL & M. PADBERG: *Polyhedral Theory/Polyhedral Computations*, in: E.L. Lawler, J.K. Lenstra, A.H.G. Rinnoy Kan, D.B. Schmoys (eds.), "The Traveling Salesman Problem", Wiley 1988, 251–360.

[32] J. HÅSTAD: *On the size of weights for threshold gates*, SIAM J. Discrete Math. **7** (1994), 484–492.

[33] M. HAIMAN: *A simple and relatively efficient triangulation of the n-cube*, Discrete Comput. Geometry **6** (1991), 287–289.

[34] F. HOLT & V. KLEE: *A proof of the strict monotone 4-step conjecture*, in: "Advances in Discrete and Computational Geometry" (B. Chazelle, J.E. Goodman, R. Pollack, eds.), *Contemporary Mathematics* **223** (1998), Amer. Math. Soc., Providence, 201–216.

[35] M. HUDELSON, V. KLEE & D. G. LARMAN: *Largest j-simplices in d-cubes: Some relatives of the Hadamard maximum determinant problem*, Linear Algebra Appl. **241–243** (1996), 519–598.

[36] R. B. HUGHES: *Minimum-cardinality triangulations of the d-cube for $d = 5$ and $d = 6$*, Discrete Math. **118** (1993), 75–118.

[37] R. B. HUGHES: *Lower bounds on cube simplexity*, Discrete Math. **133** (1994), 123–138.

[38] R. B. HUGHES & M. R. ANDERSON: *A triangulation of the 6-cube with* 308 *simplices*, Discrete Math. **117** (1993), 253–256.

[39] R. B. HUGHES & M. R. ANDERSON: *Simplexity of the cube*, Discrete Math. **158** (1996), 99–150.

[40] J. KAHN, J. KOMLÓS & E. SZEMERÉDI: *On the probability that a random ±1-matrix is singular*, J. Amer Math. Soc. **8** (1995), 223–240.

[41] V. KAIBEL & M. WOLFF: *Simple 0/1-polytopes*, European J. Combinatorics **21** (2000), 139–144.

[42] J. KOMLÓS: *On the determinant of $(0, 1)$-matrices*, Studia Sci. Math. Hungarica 2 (1967), 7–21.

[43] U. KORTENKAMP, J. RICHTER-GEBERT, A. SARANGARAJAN & G. M. ZIEGLER: *Extremal properties of 0/1-polytopes*, Discrete Comput. Geometry **17** (1997), 439–448.

[44] N. METROPOLIS & P. R. STEIN: *On a class of $(0,1)$ matrices with vanishing determinants*, J. Combinatorial Theory **3** (1967), 191–198.

[45] M. MIHAIL: *On the expansion of combinatorial polytopes*, in: Proc. MFCS'92, pp. 37–49.

[46] A. MUKHOPADHYAY: *On the probability that the determinant of an $n \times n$ matrix over a finite field vanishes*, Discrete Math. 51 (1984), 311–315.

[47] S. MUROGA: *Threshold Logic and its Applications*, Wiley-Interscience, New York 1971.

[48] D. NADDEF: *The Hirsch conjecture is true for $(0,1)$-polytopes*, Math. Programming **45** (1989), 109–110.

[49] M. G. NEUBAUER & A. J. RADCLIFFE: *The maximum determinant of ± 1 matrices*, Linear Algebra Appl. **257** (1997), 289–306.

[50] A. M. ODLYZKO: *On subspaces spanned by random selections of ± 1vectors*, J. Combinatorial Theory, Ser. A **47** (1988), 124–133.

[51] D. RIEDEL: *Berechnung des Chvátal-Gomory-Rangs von Polyedern*, Diplomarbeit, TU Berlin 1999, 52 Seiten.

[52] F. P. SANTOS: *Applications of the polyhedral Cayley trick to triangulations of polytopes*, Extended Abstract, Kotor conference, 1998.

[53] A. SCHRIJVER: *Theory of Linear and Integer Programming*, Wiley-Interscience Series in Discrete Mathematics and Optimization, John Wiley & Sons, Chichester New York 1986.

[54] W. SMITH: *Studies in Computational Geometry Motivated by Mesh Generation*, Ph.D. Dissertation, Program in Applied and Computational Mathematics, Princeton University 1988.

[55] W. SMITH: *A lower bound for the simplexity of the N-cube via hyperbolic volumes*, European J. Combinatorics **21** (2000), 131–137.

[56] GY. SZEKERES & P. TURÁN: *Extremal problems for determinants*, in Hungarian in *Math. és Term. Tud. Értesítő* (1940), 95–105; in English with a Note by M. Simonovits in "Collected Papers of Paul Turán" (P. Erdős, ed.), Vol. 1, Akadémiai Kiadó, Budapest 1990, 81–89.

[57] J. WILLIAMSON: *Determinants whose elements are 0 or 1*, Amer. Math. Monthly **53** (1946), 427–434.

[58] G. M. ZIEGLER: *Lectures on Polytopes*, Graduate Texts in Mathematics **152**, Springer-Verlag New York Berlin Heidelberg 1995, 1998; *Updates, corrections, and more* at http://www.math.tu-berlin.de/~ziegler

Günter M. Ziegler
Dept. Mathematics, MA 7-1
Technische Universität Berlin
10623 Berlin, Germany
ziegler@math.tu-berlin.de

polymake: a Framework for Analyzing Convex Polytopes

Ewgenij Gawrilow and Michael Joswig

Abstract. polymake is a software tool designed for the algorithmic treatment of polytopes and polyhedra. We give an overview of the functionality as well as of the structure. This paper can be seen as a first approximation to a polymake handbook.

The tutorial starts with the very basics and ends up with a few polymake applications to research problems. Then we present the main features of the system including the interfaces to other software products.

polymake is free software; it is available on the Internet at

http://www.math.tu-berlin.de/diskregeom/polymake/.

1. Introduction

polymake is a software package which supports research in all areas of polytope theory. Moreover, it is also used for university level education.

The key idea is the following. The user defines a polytope (or unbounded polyhedron) in one of several possible ways. This definition is stored in a file which represents the polytope. Then the user can immediately start to ask about properties of his or her polytope which are answered by polymake regardless of the initial definition. Similarly, the specified polytope can be used for constructing other polytopes. A typical polymake session could be: Specify polytope P as the convex hull of finitely many points in $\mathbb{R}^8$ and the polytope P' in terms of inequalities, also in $\mathbb{R}^8$. Define P'' to be the intersection of P and P'. Visualize the projection of P'' onto the third, fourth, and eighth coordinate.

There are very many tools available which allow to deal with various aspects of polytope theory on a computer. Among these are Avis' LRS [4], Fukuda's CDD [17], Christof and Loebel's PORTA [11], The Geometry Center's GEOMVIEW [27], and many more. The purpose of polymake is to combine the features of these (and other) programs and to go beyond. Many features of the polymake system are realized by interfacing to these programs. For an overview over existing software systems in the area of Computational Geometry see Amenta [2].

Here we describe polymake Version 1.3, released Feb 12, 1999.

2. Exploring polymake

2.1. First steps

Suppose you have a finite set of points in some vector space $\mathbb{R}^d$. Their convex hull is a polytope P. Now you want to know how many facets the polytope P has. As an example let the set S consist of the points $(0,0,0)$, $(0,0,1)$, $(0,1,0)$, $(0,1,1)$, $(1,0,0)$, $(1,0,1)$, $(1,1,0)$, $(1,1,1)$ in $\mathbb{R}^3$. Clearly, S is the set of vertices of a cube. Produce a text file **cube.poly** containing the following information.

```
POINTS
1 0 0 0
1 0 0 1
1 0 1 0
1 0 1 1
1 1 0 0
1 1 0 1
1 1 1 0
1 1 1 1
```

We have the keyword POINTS followed by eight lines, where each line contains the coordinates of one point. The coordinates are preceded by an additional '1' in the first column; the reason will be discussed later. The solution of the initial problem concerning the number of facets can now be left to **polymake**.

```
> polymake cube.poly N_FACETS
```

While **polymake** is searching for the answer, let us recall what **polymake** has to do in order to obtain the solution. It has to compute the convex hull of the given point set in terms of an explicit list of the facet describing inequalities. Then they have to be counted, which is, of course, the easy part. Nonetheless, **polymake** decides on its own what has to be done, before the final — admittedly trivial — task can be performed. In the meantime, **polymake** is done. Here is the answer.

```
N_FACETS
6
```

Depending on the individual software configuration **polymake** chooses one of the convex hull computing algorithms that have a **polymake** interface. In the previous example **polymake** might have called PORTA [17] tacitly. It is possible to specify which program is to be called, if more than one are installed.

A remark on **polymake**'s file format seems appropriate. The current version 1.3 still uses an ASCII format. The file is subdivided into several sections each of which reflects some property of the polytope. Sections are divided by blank lines. Each section starts with the (capitalized) section header. Future versions will have a binary file format in order to speed up Input/Output.

In the beginning the user defines the polytope by producing a file containing a single section (typically either POINTS or INEQUALITIES). Each subsequent call to **polymake** reads this file and looks for the desired information. If it is not there, then **polymake** searches its rule base for some sequence of computations which finally give an answer, cf. Section 5. All the relevant information which is gathered during these computations is stored into sections of the polytope file.

As a matter of fact, polymake knows a bit about standard constructions of certain polytopes. So you will never actually have to type in your 3-cube example. You can use the following command instead.

```
> cube cube.poly 3 0
```

But let us try something else. How does a typical polytope look like, which is the convex hull of 20 points on the unit sphere in $\mathbb{R}^3$? This requires one command to produce a polymake description of such a polytope and a second one to trigger the visualization. Again there is an implicit convex hull computation. But, for technical reasons[1], this time it is also necessary to determine the full combinatorial information: It has to be known which vertex is contained in which facet.

```
> rand_sphere random.poly 3 20
> polymake random.poly VISUAL
```

polymake's standard tool for visualization is GEOMVIEW [27]. It is fully interactive. For instance, it allows you to rotate and scale the aspect of your polytope. Here are two different postscript snapshots of the 1-skeleton, cf. Figure 1.

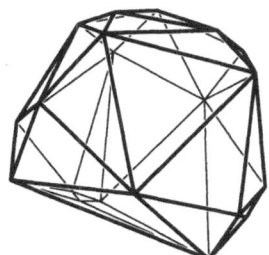 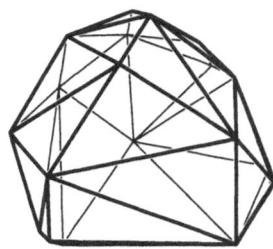

FIGURE 1. Convex hull of 20 points chosen uniformly at random on the unit sphere in $\mathbb{R}^3$. Two views of the same polytope.

Polytopes most naturally appear as solution sets of linear programs. Consider the set of inequalities below on the left.

$$
\begin{array}{rcl}
x & \geq & 0 \\
y & \geq & 0 \\
1 - x - y & \geq & 0 \\
52 - 2x - 3y & \geq & 0
\end{array}
\qquad
\begin{array}{l}
\text{INEQUALITIES} \\
0 \quad 1 \quad 0 \\
0 \quad 0 \quad 1 \\
1 \ -1 \ -1 \\
52 \ -2 \ -3
\end{array}
$$

In polymake's syntax this could be phrased as indicated above on the right.

People working in optimization usually prefer other file formats (such as LP), where it is also possible to keep the names of the variables. polymake is aware of

---

[1] GEOMVIEW's OOGL files contain the vertices and the vertex facet incidence information.

this. LP format files can be converted to `polymake` format and conversely. The variable names are kept, but `polymake` does not use them.

It is not difficult to see that the above system of inequalities defines a triangle with vertices $(0,0)$, $(1,0)$, $(0,1)$. In particular, the last of the input inequalities is redundant. Assume the file `linear_program.poly` contains the inequalities listed above.

```
> polymake linear_program.poly VERTICES N_FACETS
VERTICES
1 0 0
1 0 1
1 1 0

N_FACETS
3
```

It is one essential feature of `polymake` that it is possible to define a polytope in several different ways: either as the convex hull of a set of points or as the intersection of half-spaces. If you are asking for a specific property of the polytope defined, the syntax of the call to `polymake` does not depend on the definition. This is entirely transparent for the user.

Suppose now you are not interested in a particular coordinate representation of a polytope. But instead you want to focus on the combinatorial properties only. `polymake` supports this point of view, too. You can specify a polytope in terms of its vertex-facet-incidence matrix. For each facet you have a line with a list of the vertices contained in that facet. The vertices are specified by numbers. They are numbered consecutively starting from 0. In each row the vertices are listed in ascending order. This is a valid `polymake` description of a square.

```
VERTICES_IN_FACETS
{0 1}
{1 2}
{2 3}
{0 3}
```

Note that in this situation `polymake` assumes that you actually specified a polytope in this way. `polymake` has (almost) no means to check this.

The dimension of a polytope, i.e. the dimension of its convex hull, is an intrinsic property. It does not depend on the coordinate representation. The vertex-facet-incidence matrix suffices for `polymake` to compute the dimension. Assume the data above was stored in a file named `square.poly`.

```
> polymake square.poly DIM
DIM
2
```

Sometimes one wants to construct new polytopes from old ones. Suppose we need a prism over our square, which clearly gives a cube. The trailing parameter `-noc` (no coordinates) in the call of prism indicates that we want a combinatorial description of the prism; the same program can also produce a coordinate description, if coordinates for the input polytope are provided.

```
> prism prism.poly square.poly -noc
```

The cube is a simple polytope, but it is not simplicial, cf. Section 4.2.

```
> polymake prism.poly SIMPLE SIMPLICIAL
SIMPLE(true)

SIMPLICIAL(false)
```

2.2. Constructing a neighborly cubical polytope

In this section we want to give an example of how polymake can be used to construct certain polytopes with particularly interesting properties. We will construct a 4-dimensional polytope whose vertex-edge-graph is isomorphic to the graph of the 5-dimensional cube. More generally, the construction can be used to produce cubical $2d$-polytopes which have the same $(d-1)$-skeleton as the $(2d+1)$-dimensional cube, cf. Section 4.2. See [23, Section 4.1] for details.

Let $Q = [-1,1]^d$ be a d-dimensional cube with ± 1 vertex coordinates. For an isomorphic cube with ± 2 coordinates write $2Q$. The direct product $Q \times 2Q$ is isomorphic to the $2d$-dimensional cube. Then the convex hull

$$P = \mathrm{conv}(Q \times 2Q \ \cup \ 2Q \times Q)$$

of the union of two combinatorial cubes is a polytope with the desired properties.

In order to understand what is going on it may be worth to visualize the case $d = 1$ where we obtain a 2-polytope whose vertex set is isomorphic to the vertex set of the 3-cube, i.e. an octagon, cf. Figure 2.

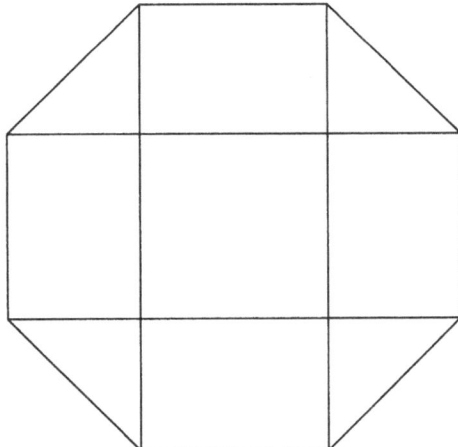

FIGURE 2. The convex hull of two rectangles yielding an octagon.

Now we use **polymake** to construct the example for $d = 2$. We start by producing two cubes, Q and $2Q$.

```
> cube Q 2 1
> cube 2Q 2 2
```

Then we can form their product in two ways and take the common convex hull.

```
> product Qx2Q Q 2Q
> product 2QxQ 2Q Q
> conv P Qx2Q 2QxQ
```

The resulting polytope can be visualized as a Schlegel diagram, cf. Figure 3. See Figure 7 for an indication how Schlegel diagrams are constructed.

```
> polymake P SCHLEGEL
```

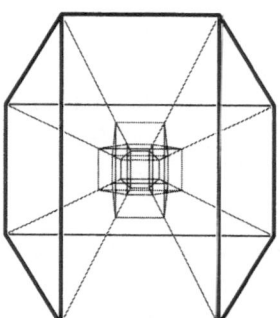 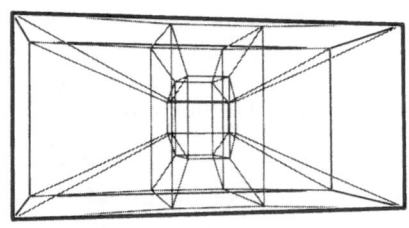

FIGURE 3. Two Schlegel diagrams of a cubical 4-polytope with the same graph as the 5-cube.

2.3. A strange 3-cube

The combinatorics of d-dimensional cubes can certainly be considered as well understood. But, when it comes to realizations of such cubes, it turns out that there are quite a few combinatorial cubes with surprising properties. The Klee-Minty and Goldfarb cubes, for instance, yield well-known examples which show that linear programming is not so easy — not even over cubes; cf. Figure 9 for the directed graph of a 4-dimensional Klee-Minty cube.

Here we discuss something else. The standard d-cube $[0, 1]^d$ has the property that opposite facets are contained in parallel hyperplanes, while non-opposite facets are perpendicular. Is there a cube with the property that each facet is perpendicular to its opposite? A direct argument shows that this is not possible for 2-cubes, that is, quadrangles. But, one can come arbitrarily close.

Günter M. Ziegler constructed a 3-cube with the property that each facet is perpendicular to its opposite, thereby answering a question of Sharir. Here is a polymake input, which describes the cube in terms of inequalities.

```
INEQUALITIES
25     -2    -25     10
-2     25      2     10
25     -2     25     10
-2     25     -2     10
 0      0     -1     -1
 2      0     -1      1
```

If numbered from 0 through 5, the hyperplane corresponding to inequality number $2i$ is clearly perpendicular to the hyperplane corresponding to inequality number $2i + 1$: Omit the first coordinates, and compute the scalar products to check. But do these inequalities define a cube such that the perpendicular inequalities are opposite facets?

```
> polymake strange-cube.poly
   N_FACETS VERTICES_IN_FACETS DIM CUBICALITY N_FACETS
6

VERTICES_IN_FACETS
{0 1 6 7}
{2 3 4 5}
{4 5 6 7}
{0 1 2 3}
{1 2 4 6}
{0 3 5 7}

DIM
3

CUBICALITY
3
```

polymake tells us that we have 6 facets, that is, there was no redundancy among the inequalities. Moreover, the incidence information reveals that facet number $2i$ and facet number $2i + 1$ do not share any vertices. But, is it a cube? Well, it is 3-dimensional, and the largest dimension in which all faces are combinatorial cubes equals 3. Here is a picture, cf. Figure 4. In order to understand how it really looks like one has to rotate the image back and forth on the screen.

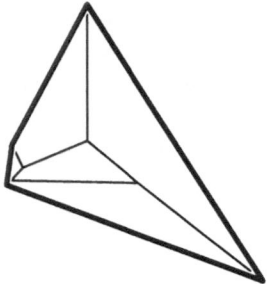

FIGURE 4. A 3-cube with pairwise perpendicular opposite facets.

3. polymake's view on polytopes

The purpose of this section is to outline the mathematical foundations of **polymake**.
We recall the definition and the basic properties of polytopes and polyhedra. For
more details see Ziegler [33].

Take any set of points $P \subset \mathbb{R}^d$. Let

$$
\mathrm{conv}(P) = \left\{ \sum_{p \in P} c_p p \ \middle|\ c_p \geq 0, \ \sum_{p \in P} c_p = 1 \right\}
$$

and

$$
\mathrm{cone}(P) = \left\{ \sum_{p \in P} c_p p \ \middle|\ c_p \geq 0 \right\}
$$

denote the *convex hull* and the *cone* of P, respectively. The convex hull of finitely
many points in $\mathbb{R}^d$ is called a *V-polytope*. The intersection of finitely many halfspaces of $\mathbb{R}^d$ is called an *H-polyhedron*. The *dimension* of a polyhedron is the
dimension of its affine hull. A d-dimensional polyhedron is usually called a *d-
polyhedron*.

For the theory of polytopes it is essential that V-polytopes and H-polyhedra
merely are two descriptions of basically the same thing: Each V-polytope is a
bounded H-polyhedron, and vice versa.

Therefore, it is not necessary to distinguish between V-polytopes and bounded
H-polyhedra. Both are called *polytopes*. The elements of the bigger class of not necessarily bounded H-polyhedra are called *polyhedra*.

polymake allows to work with polytopes as well as unbounded polyhedra.
However, when it comes to combinatorics we prefer polytopes. See the details
below. Specifying a V-polytope is done by producing a file which contains a section POINTS which has one line per point. Each point is given as a row vector of
signed homogeneous coordinates, cf. Figure 5. Similarly, an H-polyhedron can be
defined in terms of INEQUALITIES. In order to encode a linear inequality, that is, a
halfspace, choose a normal vector of the bounding hyperplane pointing inside the

halfspace. Its signed homogeneous coordinates (as a column vector) encode the inequality. Now the incidence between a point and a hyperplane is signaled by the vanishing of the standard scalar product of the two vectors.

The equivalence of V-polytopes and bounded H-polyhedra is only half the truth. The two ways to describe polytopes are, in principal, equivalent, but one of them can be much more efficient than the other. For instance, the d-cube has 2^d vertices but only $2d$ facets. A standard procedure for the mutual conversion is provided by the Fourier-Motzkin elimination method, see Ziegler [33, Section 1.2]. For an overview about common convex hull computation strategies see Edelsbrunner [14, Chapter 8]. See also Section 4.1.

Working with mathematical objects on an algorithmic level is often made easier if unique representations of these objects can be obtained. For a polytope P a normal form is immediately at hand. Among all sets $S \in \mathbb{R}^d$ with $\text{conv}(S) = P$ choose the unique minimal one (with respect to inclusion or cardinality): the set of *vertices*. These are the extremal points of P, which cannot be written as a convex combination of other points of P. For arbitrary polyhedra we have the following.

Let $P \subset \mathbb{R}^d$ be a polyhedron. Then there are finite sets $\{v_1, \dots, v_k\} \subset P$, $\{w_1, \dots, w_l\} \subset \mathbb{R}^d$ such that

$$P = \text{conv}(v_1, \dots, v_k) + \text{cone}(w_1, \dots, w_l).$$

Unfortunately, this does not lead to a normal form because for an unbounded polyhedron there may be several different sets generating the polyhedron of the same cardinality. Take any affine line L in $\mathbb{R}^d$. This is clearly a polyhedron as it is the intersection of $d - 1$ hyperplanes in general position. Every hyperplane is the intersection of two halfspaces. Choose two points $x, y \in L$. Then $L = \text{conv}(x) + \text{cone}(x - y, y - x)$. Similarly for any other pair of points on L. There is no natural choice for a specific pair. An examination of the problem shows that this is about the only bad thing that can actually happen. If we exclude this case, then we can make a stronger statement than above.

Let $P \subset \mathbb{R}^d$ be a polyhedron which does not contain an affine line. Then there are (essentially) *unique* finite sets $\{v_1, \dots, v_k\} \subset P$, $\{w_1, \dots, w_l\} \subset \mathbb{R}^d$ of minimal cardinality such that $P = \text{conv}(v_1, \dots, v_k) + \text{cone}(w_1, \dots, w_l)$. Here $\text{cone}(\emptyset) = \{0\}$. The vectors w_i are called *rays*. They are unique only up to a positive multiple.

Now let $P = \text{conv}(v_1, \dots, v_k) + \text{cone}(w_1, \dots, w_l)$ be a *pointed polyhedron*, that is, a polyhedron which does not contain an affine line. Consider an embedding ι of $\mathbb{R}^d$ into the set of rays in $\mathbb{R}^{d+1}$. We set $\iota(v_i) = \mathbb{R}_{\geq 0}(1, v_i)$. Now there is a canonical extension of ι to the set of rays of $\mathbb{R}^d$ mapping w_i to $\mathbb{R}_{\geq 0}(0, w_i)$. Using signed homogeneous coordinates along these lines makes it widely unnecessary to distinguish between vertices and rays. This is the coordinate model which is used by **polymake**, cf. Figure 5. Note that we do not force the vertices (and rays) to appear in a particular order, thus failing to arrive at a normal form for pointed polyhedra in a stricter sense. Vertices and rays are stored in the **polymake** section

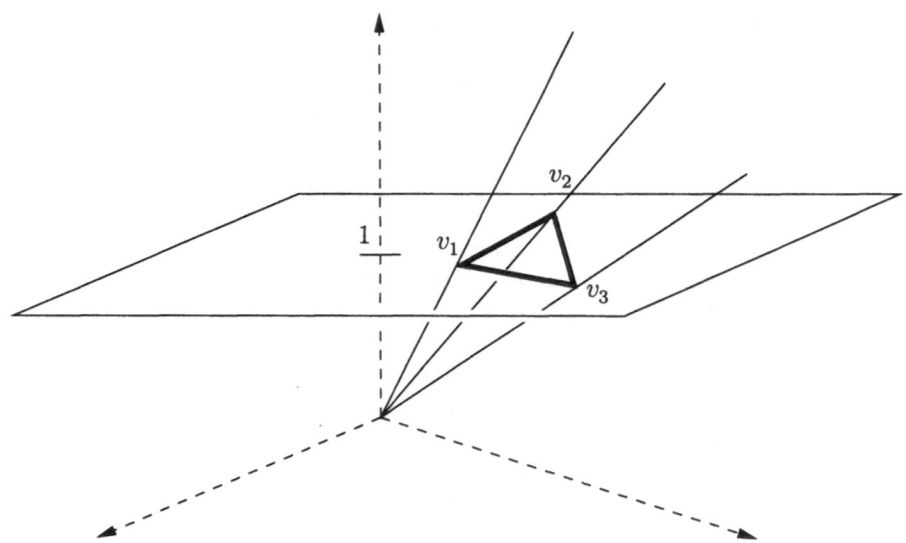

FIGURE 5. Signed homogeneous coordinate model. Geometrically
we pass from the original polyhedron to a cone.

VERTICES, vertices can be recognized by a leading positive coordinate. Negative
leading coordinates are forbidden.

The *signed* homogeneous coordinate model is very similar to the concept of
(unsigned) homogeneous coordinates used in projective geometry. However, the key
difference is that in projective geometry only incidence matters, while in convex
geometry questions of orientation additionally come into play.

In what follows we discuss the geometric properties of full-dimensional pointed polyhedra in $\mathbb{R}^d$. For this it is best to think of such a polyhedron turned into
a cone as in Figure 5 which then is intersected with the unit sphere $\mathbb{S}^d$ in $\mathbb{R}^{d+1}$.
Each affine k-space in $\mathbb{R}^d$ becomes a k-dimensional sphere embedded in $\mathbb{S}^d$. A full-
dimensional polytope becomes a *chamber* of a spherical hyperplane arrangement.
In fact, a polytope corresponds to two (opposite) chambers, and the signed homogeneous coordinate model makes a decision in favor of one of them, cf. Figure 6.
We refer to this view as the *spherical model* of a polytope.

The distinction between the two opposite chambers is geometrically achieved
by introducing one special hyperplane, which sometimes is called the *far hyperplane F*. The far hyperplane contains (the images of) all the rays, that is, in our
model

$$F = \left\{ (x_0, \dots, x_d) \in \mathbb{R}^{d+1} \mid x_0 = 0 \right\}.$$

Suppose you are given the set of facets of a full-dimensional polyhedron together
with the far hyperplane in the spherical model. Then the property of being pointed

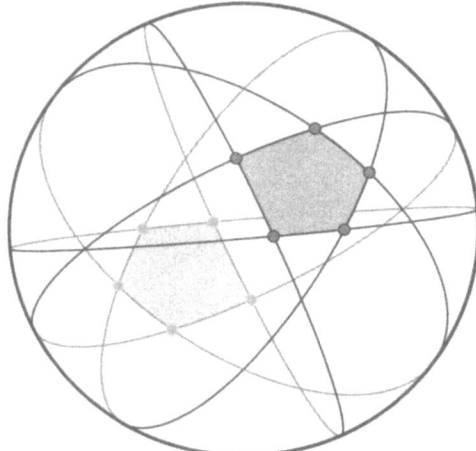

FIGURE 6. A 2-polytope in the spherical model, that is, a hyperplane arrangement on the 2-sphere with one chamber (and its opposite) shaded. The far hyperplane is the boundary of the projection. Cinderella [30] postscript output.

is equivalent to the property that all chambers of the spherical hyperplane arrangement are polytopal. In particular, except for the choice of the far hyperplane, there is no further distinction between a polytope and a pointed polyhedron in the spherical model. The far hyperplane F partitions the sphere into two open hemispheres H^+ and H^- and the intersection of F with the sphere itself. The Euclidean space, where the polyhedron naturally lives in, becomes the open hemisphere H^+. Its topological closure is the closed hemisphere $\overline{H^+}$. Further we do not distinguish between the vertices and the rays of a pointed polyhedron.

We want to investigate the spherical model and the relationship between polytopes and pointed polyhedra a little bit further. Linear transformations of $\mathbb{R}^{d+1}$ induce *projective linear transformations* of the d-dimensional projective space (whose points are the 1-dimensional subspaces of $\mathbb{R}^{d+1}$, that is, the lines through the origin). The projective linear transformations which fix the far hyperplane are precisely the affine transformations. In the following we are looking for a projective linear map which transforms a given pointed polyhedron into a polytope. As we are interested in doing this effectively, we must be a bit more explicit. For *convex* geometry some care is needed in applying projective transformations, which are not affine.

In order to illustrate these difficulties consider, for example, the triangle

$$T = \text{conv}((0,0), (1,0), (0,1))$$

which becomes the cone

$$C(T) = \text{cone}((1,0,0),(1,1,0),(1,0,1))$$

in our model. Points and rays are row vectors here, so linear transformations will be applied from the right. Suppose we want to apply to $C(T)$ the linear transformation

$$\begin{pmatrix} 1 & 0 & 0 \\ -2 & -1 & 0 \\ -2 & 0 & -1 \end{pmatrix}$$

In the signed homogeneous coordinate model the rays of $C(T)$ would be mapped as follows:

$$\begin{aligned}
\mathbb{R}_{\geq 0}(1,0,0) &\mapsto \mathbb{R}_{\geq 0}(1,0,0), \\
\mathbb{R}_{\geq 0}(1,1,0) &\mapsto \mathbb{R}_{\geq 0}(-1,-1,0), \\
\mathbb{R}_{\geq 0}(1,0,1) &\mapsto \mathbb{R}_{\geq 0}(-1,0,-1).
\end{aligned}$$

Unfortunately, two of the images are not contained in $\overline{H^+}$. If we try to remedy the situation by skipping the sign restriction, that is, we try to interpret the result in the projective space, then something even stranger happens. Now all the three rays are fixed (because $\mathbb{R}_{\geq 0}(-1,-1,0) = \mathbb{R}(1,1,0)$ and, similarly, $\mathbb{R}_{\geq 0}(-1,0,-1) = \mathbb{R}(1,0,1)$), but the point $(1/3,1/3)$, which corresponds to the ray $\mathbb{R}_{\geq 0}(1,1/3,1/3)$, is mapped to $\mathbb{R}_{\geq 0}(-1,-1,-1)$, which then is the same as $\mathbb{R}(1,1,1)$. But, $(1/3,1/3)$ is an interior point of T while $(1,1)$ is not.

This discussion leads us to one more definition. A linear transformation of $\mathbb{R}^{d+1}$ is called *admissible* for a pointed polyhedron $P \subset \overline{H^+}$ in the spherical model $\mathbb{S}^d$ if for all points $p \in P$ we have $\lambda(p) \in \overline{H^+}$ and $\lambda(P)$ is the conic hull of the image of the vertices/rays of P. Note that each affine transformation is admissible for every pointed polyhedron.

Given a pointed polyhedron P we want to exhibit a sequence of admissible transformations which transform P into a polytope. The following lemma does not require the spherical model as it deals with affine transformations only.

Lemma 3.1. *Every pointed polyhedron P is affinely equivalent to a (pointed) polyhedron which is contained in the positive orthant.*

Proof. Because P is pointed, it has at least one vertex. Translate any such vertex into the origin. We can assume that P is full-dimensional, i.e. the points of P span $\mathbb{R}^d$ as a vector space. This implies that we can select a basis $(F_1, \dots, F_d)$ for the dual space $(\mathbb{R}^d)^*$ from the facets through the origin. There is a linear transformation λ which maps each facet F_i into the hyperplane $\{(x_1, \dots, x_d) \mid x_i = 0\}$. $\square$

A polyhedron which is contained in the positive orthant cannot contain any affine line, that is, it is pointed. Now we employ our spherical model.

Proposition 3.2. *There is a map which transforms every (pointed) polyhedron contained in the positive orthant into a bounded polytope. Moreover, this transformation is admissible for every such polyhedron.*

Proof. Suppose that P is a polyhedron contained in the positive orthant. We use the embedding $\iota : (x_1, \ldots, x_d) \mapsto \mathbb{R}_{\geq 0}(1, x_1, \ldots, x_d)$ of the affine d-space into its spherical closure $\overline{H^+}$. Now the transformation π given by the $(d+1) \times (d+1)$-matrix

$$
\begin{pmatrix}
1 & 0 & \cdots & \cdots & 0 \\
1 & 1 & \ddots & & \vdots \\
1 & 0 & \ddots & \ddots & \vdots \\
\vdots & \vdots & \ddots & \ddots & 0 \\
1 & 0 & \cdots & 0 & 1
\end{pmatrix}
$$

fixes the origin, and it maps the positive orthant into a d-simplex, namely the pyramid over the $(d-1)$-simplex spanned by ι-images of the unit vectors of $\mathbb{R}^d$ (with the origin as the apex).

The admissibility of π to any polyhedron in the positive orthant follows from the fact that π only has non-negative entries. $\square$

In particular, this implies that every pointed polyhedron is combinatorially equivalent to a polytope, if one takes into account the far faces. This is what polymake does. For instance, polymake provides a function which computes the diameter of (the vertex-edge graph of) a pointed polyhedron. Far vertices (= rays) and edges between them do play a role. Note that this point of view may differ from those of other authors.

If the far face is a facet then the application of π is combinatorially equivalent to bounding the polyhedron with a hyperplane beyond all the vertices.

The transformation steps in Lemma 3.1 and Proposition 3.2 can be made explicit within the polymake system by calling the clients orthantify and bound, respectively, cf. Section 5.4.

4. Features

This is an overview of what can be done with polymake. We want to stress that many of the most important functions rely on implementations by other research groups. polymake provides a variety of interfaces for these.

4.1. Convex hull computation

polymake does not have own means for a convex hull computation. Instead we rely on a variety of free software packages. The current version 1.3 of polymake offers interfaces to the programs listed below.

See Ziegler [33, Section 1.2] for an introduction to convex hull computation by Fourier-Motzkin elimination. See also Avis, Bremner and Seidel [5] as well as

Bremner [10] for a thorough discussion of various convex hull computation methods and their complexity.

4.1.1. CDD   This package has been developed by Fukuda. It implements the double description (or Fourier-Motzkin elimination) method to compute the convex hull of a full-dimensional polyhedron. The program comes in two flavors, one using exact rational arithmetic, the other one using floating-point arithmetic. Both are accessible through `polymake`.

Since Version 0.76 CDD uses the GNU Multiprecision Library (GMP) via `polymake`'s C++-wrappers. See also the Section 5.7 on matters of arithmetic in `polymake`.

For details see Fukuda's CDD homepage [17] as well as Fukuda and Prodon [18].

4.1.2. LRS   The algorithm behind LRS is the reverse search algorithm of Avis and Fukuda [6], implemented by Avis [4, 3]. Like CDD it is also restricted to full-dimensional input.

4.1.3. PORTA   This is Christof's and Loebel's implementation of the convex hull computation by Fourier-Motzkin elimination. PORTA [11] does not require full-dimensional input.

4.2. Combinatorics

`polymake` is able to work with polytopes on a combinatorial level. Many interesting properties of a polytope do not depend on the particular coordinate representation but rather on the face lattice of the polytope. `polymake` allows the user to specify the combinatorial type of a polytope in terms of its vertex-facet-incidence matrix.

Some caution is necessary here because `polymake` does not provide means to check whether an incidence matrix actually describes a polytope or not. `polymake` can detect some of the most blatant errors; in this case the user will be informed. But it is also possible to feed `polymake` with an incidence matrix which contains a much more subtle error. Asking `polymake` about properties of this non-polytope might, of course, yield some answers, which are nonsense. It might fail on some later computation with a rather bizarre error message[2].

Concerning the combinatorics of a polytope we decided to slightly deviate from `polymake`'s overall design principles. While `polymake` does provide the functionality to compute the face lattice of a polytope, this information is not stored into the polytope file after its first computation. Instead it is computed from scratch (and displayed on the screen) each time the user asks for it. We took this approach in order to avoid problems arising from the sheer size of the face lattice.

For further information that is typically derived from the face lattice we prefer to work with a suitable abstraction. It turns out that many combinatorial properties such as simplicity, simpliciality, neighborliness, etc. only depend on the flag vector of the polytope, cf. Bayer and Lee [7, Section 3.7]. More precisely, even

---

[2]This is due to the fact that `polymake` assumes that the information within one file is consistent. Some of the algorithms rely on this consistency without checking.

the $f^{(2)}$-*vector* suffices. We define $f_{i,k}^{(2)}(P) = f_{k,i}^{(2)}(P)$ as the number of incidences between i-dimensional and k-dimensional faces of the polytope P. So, the $f^{(2)}$-vector is a restriction of the usual flag vector to pairs of dimensions. It contains the f-vector on the diagonal. We abbreviate $f_k(P) = f_{k,k}^{(2)}(P)$.

A polytope is *simplicial* if all its proper faces are simplices. From the $f^{(2)}$-vector it can easily recognized whether a polytope has this property.

Lemma 4.1. *The d-polytope P is simplicial if and only if* $f_{0,d-1}^{(2)}(P) = d f_{d-1}(P)$.

An example for a simplicial 3-polytope can be seen in Figure 1.

```
> polymake random.poly SIMPLICIAL
SIMPLICIAL(true)
```

Recently, there seems to be a growing interest in *cubical* polytopes, i.e. polytopes whose proper faces are (combinatorial) cubes, cf. Blind and Blind [8] [9], see also the references in [23]. Given that the graph of a cubical polytope does not contain a triangle [22, Proposition 3.2], the following characterization follows from an easy induction on the dimension.

Lemma 4.2. *The d-polytope P is cubical if and only if* $f_{0,k}^{(2)}(P) = 2^k f_k(P)$ *for all $k < d$ and the vertex-edge-graph does not contain any triangles.*

The polytope P in Figure 3 is cubical.

```
> polymake P CUBICAL
CUBICAL(true)
```

Other combinatorial invariants which are accessible via **polymake** include the vertex-edge-graph, the vertex degrees, the Altshuler determinant and more. As already discussed in Section 3, **polymake** considers a pointed polyhedron as combinatorially equivalent to a polytope.

4.3. Linear optimization

polymake does not have a direct interface to any linear program solver. But, **polymake** provides conversion between LP format and **polymake** format. Use the Perl script lp2poly that comes with the **polymake** distribution.

Moreover, **polymake** can solve (small) linear programs on its own. We have implemented a geometer's version of the simplex algorithm, which works as follows. From a given set of inequalities compute the vertices by a dual convex hull computation, cf. Section 4.1. Then determine the combinatorial structure, in particular, the vertex-edge-graph of the polyhedron. When all this is known, one can start from an arbitrary vertex and move to an adjacent better vertex as long as possible. As of version 1.3, only the Maximum Increase pivot rule is implemented, but future versions of **polymake** will have a more sophisticated tool box to study various pivot strategies.

The algorithm indicated clearly does not allow for efficient linear programming but it has the advantage that it can also deal with *abstract objective functions* as studied by Kalai [24] and others.

4.4. Visualization

Up to now `polymake` does not provide many tools for visualization on its own. Instead, once again, we usually rely on packages developed by other groups.

4.4.1. GEOMVIEW Most naturally, visualization starts in two and three dimensions. To get an impression of the overall shape of a 3-polytope one wants to see either a solid or a wire model. The program GEOMVIEW [27] can do both.

GEOMVIEW is run through `polymake` directly. In the introduction you already saw the call

```
> polymake random.poly VISUAL
```

which produced the image in Figure 1. Because GEOMVIEW is the default tool for visualization the preceding call is equivalent to

```
> polymake random.poly VISUAL geomview
```

Visualization of arbitrary higher-dimensional objects is clearly more difficult. One common general method of four-dimensional visualization is the animation of three-dimensional slices of the object to be visualized. For 4-polytopes there is another well-known method: the visualization by Schlegel diagrams. It has the advantage that the whole polytope can be visualized in a single image. Moreover, all facets and their relative positions are correctly displayed. The idea is to project the 4-polytope onto one of its facets. One obtains a polytopal complex which, essentially, has the same combinatorial properties as the boundary of the polytope.

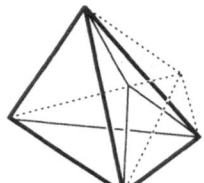  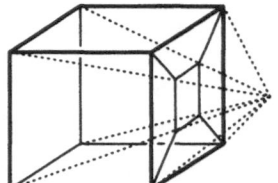 

FIGURE 7. Construction of Schlegel diagrams of the tetrahedron and the 3-cube. These pictures are taken from [33, pp. 134/135].

In general, the polytopal complex (and the image) depend on the facet chosen, see, e.g. the different Schlegel diagrams of the permutahedron in Figure 8. For a thorough discussion of Schlegel diagrams see [33, Chapter 5].

4.4.2. GRAPHLET The visualization of objects which live in dimensions even higher than 4 is not entirely futile. `polymake` offers an interface to the graph drawing/visualizing tool GRAPHLET [16]. This way it is still possible to see the relative positions of the vertices in the graph. More information can be obtained if one characterizes specific faces by means of linear (or abstract) objective functions. `polymake` then passes information about the minimal and the maximal vertices to GRAPHLET. All the vertices of the minimal face are displayed in blue, while the

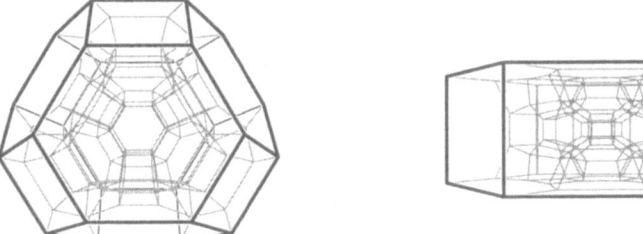

FIGURE 8. Two Schlegel diagrams of the 4-permutahedron.

maximal vertices are red. It is also possible to assign colors to all the vertices such that the same color indicates the same level with respect to the specified linear function.

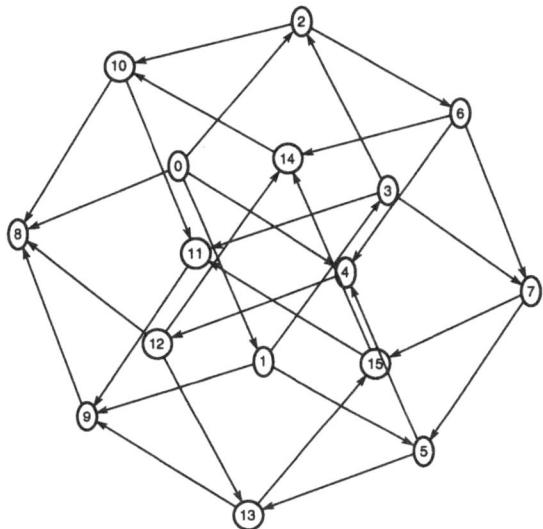

FIGURE 9. Directed graph of the 4-dimensional Klee-Minty cube. There is an ascending path through all the vertices from the bottom vertex 0 to the top vertex 8.

Of course, this interface also allows to investigate the *dual* graph of a polytope, that is, the facet-ridge graph, which is the same as the vertex-edge graph of the dual polytope.

4.4.3. GALE DIAGRAMS   There is a convenient way to visualize polytopes with few vertices: The *(affine) Gale diagram* of a d-polytope with n vertices is an $(n-d-2)$-dimensional point configuration. polymake can produce direct Postscript output for 1- and 2-dimensional affine Gale diagrams. In the 1-dimensional case it also generates the linear 2-dimensional picture; see Figures 10 and 11. For details about Gale transforms and Gale diagrams see Ziegler [33, Section 6.4].

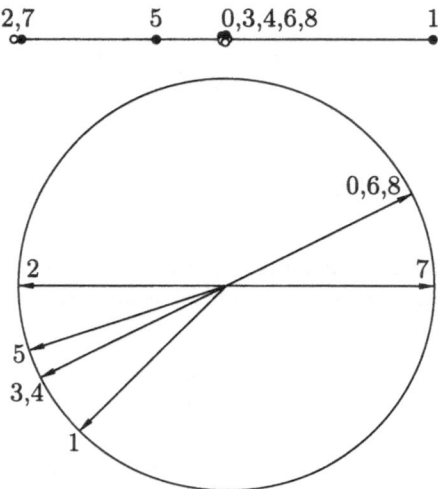

FIGURE 10. Affine and linear Gale diagrams of a 6-dimensional 0/1-polytope with 9 vertices.

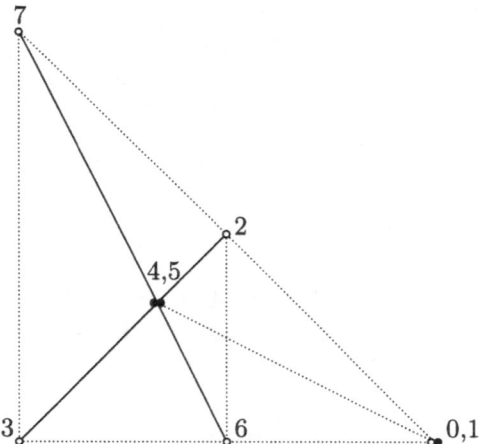

FIGURE 11. Affine Gale diagram of a 4-dimensional 0/1-polytope with 8 vertices.

As can be seen from the Figures 10 and 11, it happens that the Gale transform maps distinct vertices of the polytope to the same vector. In this case the corresponding vector of the linear Gale diagram is drawn only once; its labeling lists all the vertices mapped there, separated by a comma. In the affine Gale diagram a point is drawn for each vertex. The images of a set of points with the same coordinates are spread out by a little in order to allow them to be distinguished. Starting on the right they are arranged in counter-clockwise order; this allows to determine which vertex gets which color. For instance, in the affine Gale diagram of Figure 10 vertex 2 is black and vertex 7 is white.

For 2-dimensional affine Gale diagrams, as in Figure 11, a solid line indicates that the Gale diagram has points on at least three different positions of this line. A dashed line is drawn between any two points of the same color.

4.4.4. POVRAY   We also offer an interface to the ray-tracer POVRAY [15]. This allows the user to produce high-quality pictures of his/her favorite 3-polytopes. Schlegel diagrams can also be visualized in this way.

This interface greatly differs from the interface to GEOMVIEW. POVRAY is not run through **polymake**. Instead **polymake** produces a file which can be used as input to POVRAY. Thus the user can call the ray-tracer with individually chosen parameters.

5. Inside polymake

In the preceding section we mainly described **polymake** as a common interface to a variety of standard implementations concerned with the algorithmic treatment of polytopes. However, this is by no means the only genuine feature. **polymake**'s strength is its ability to choose on its own a way to obtain answers to the user's questions about some polytope. Moreover, **polymake** offers several ways to extend the system. In this section we want to give some indication about the general concepts.

The overall structure is sketched in Figure 12. The **polymake** server consists of two parts, the **polymake** engine and its rule base. The functionality of the interfaces to other software packages have already been described among the features, cf. Section 4; in particular, see Sections 4.1 and 4.4.

The 'mathematical knowledge' about polytopes and polyhedra is contained in the rule base, the clients (and, of course, the interfaced programs). In particular, the engine does not know anything about polytopes. Or, to phrase it differently, it is well conceivable to build a different rule base around the **polymake** engine in order to obtain a similar functionality in an entirely different context.

5.1. Two examples

Let us recall from Section 2 that the user is allowed to describe a polyhedron in its H-representation or its V-representation. Either way this does not impose any restriction on the questions the user is allowed to ask. Let us have a closer look at two further examples.

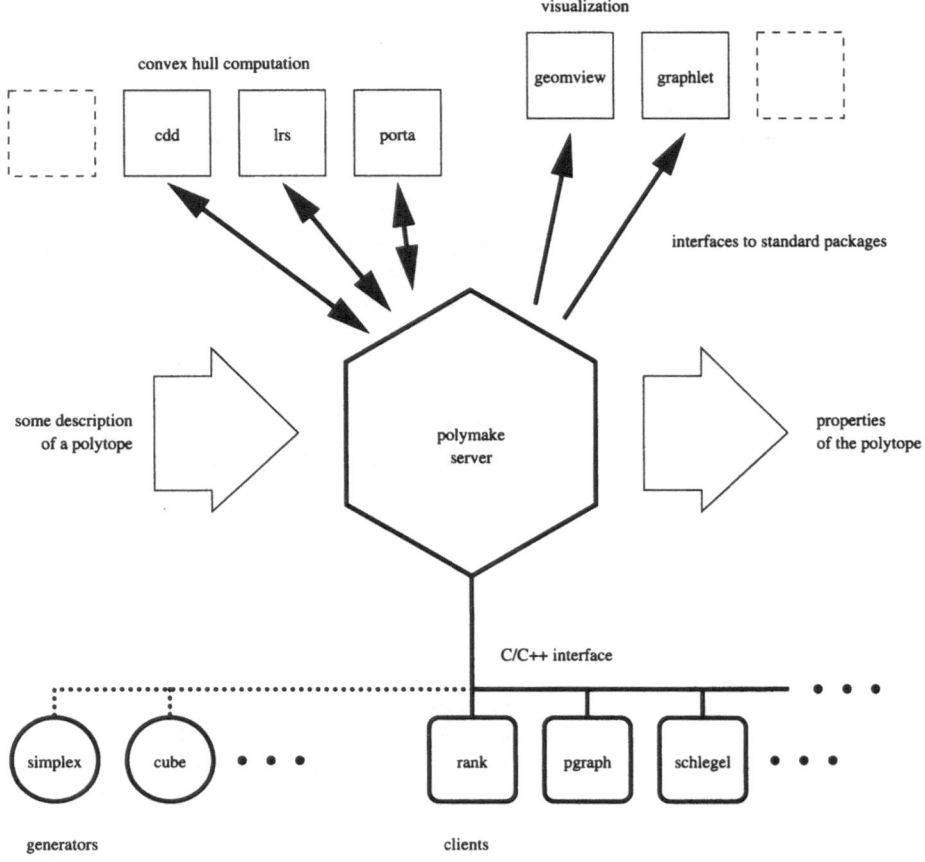

FIGURE 12. A schematic overview of **polymake**'s structure.

Assume we have some V-polytope represented by the file **v.poly** containing the section **VERTICES** with the coordinate description of all vertices, and we have a (bounded) H-polyhedron represented by the file **f.poly** containing the section **FACETS** with the coordinate description of all facets. Suppose now we want to know the vertex barycenter of each polytope. **polymake** has a section **VERTEX_BARYCENTER**, and it is possible to obtain the desired information from **polymake** through a command line call similar to the ones described above. Specifying the additional flag **-n** in the **polymake** call reveals what **polymake** is planning to do in order to satisfy this request. No actual computation is performed in this case.

```
> polymake -n v.poly VERTEX_BARYCENTER
polymake: reading rules from ... [several lines skipped here]
polymake: composing a minimum weight rule chain... spent 0 sec.
```

```
polymake: applying rule 'BOUNDED(true|false) : VERTICES(*) | POINTS(*)'
polymake: applying rule 'VERTEX_BARYCENTER : BOUNDED(true), VERTICES(...)'
```

After telling the user where `polymake` actually draws its knowledge from it says the following. Firstly, it should be checked whether the input polyhedron is actually bounded. If this holds true, then the barycenter can be computed directly from the vertices. We have quite a different situation in the other case.

```
> polymake -n f.poly VERTEX_BARYCENTER
polymake: reading rules from ... [several lines skipped here]
polymake: composing a minimum weight rule chain... spent 0 sec.
polymake: applying rule 'AMBIENT_DIM : FACETS(*) | INEQUALITIES(*)'
polymake: applying rule 'REL_INT_POINT, AFFINE_HULL : FACETS | ...
polymake: applying rule '?porta, VERTICES, N_VERTICES : ...
polymake: applying rule 'BOUNDED(true|false) : VERTICES(*) | POINTS(*)'
polymake: applying rule 'VERTEX_BARYCENTER : BOUNDED(true), VERTICES(...)'
```

The last two lines are exactly the same as before. But this time `polymake` needs three more steps in order to obtain the V-representation from the H-representation of the polytope. It suggests that PORTA should be called to handle this, but PORTA needs a point in the relative interior as an additional input. So this has to be done at the beginning. The very first rule appears for technical reasons.

5.2. The rule base

The rules specify which properties can be computed directly from which other properties. Each rule consists of the *header*, describing the functionality, and the *body*, implementing the functionality. Syntactically, the header of a rule which computes the properties o_1, o_2 to o_m from the input i_1, i_2 to i_n appears as

$$o_1, o_2, \ldots, o_m : i_1, i_2, \ldots, i_n$$

The rule body essentially is a piece of Perl code; cf. Wall, Christiansen and Schwartz [32] for an introduction to the programming language Perl. From the point of view of the engine rules are atomic.

Basically, there are three types of rules, which can be distinguished by the complexity of the implemented functionality.

5.2.1. SIMPLE RULES They do something which can be done in very few lines of Perl code. An example is the rule

```
AMBIENT_DIM : FACETS(*) | INEQUALITIES(*)
($_) = give("FACETS(*)[0] | INEQAULITIES(*)[0]");
my $d=split;
take("AMBIENT_DIM", $d-1)
```

which does nothing more than counting the length of a coordinate vector. Note the vertical bar "|" on the right hand side of the rule header. This indicates that the input are either **FACETS** or **INEQUALITIES**, where **FACETS** are preferred if both are available. Throughout this paper we avoid to discuss the ramifications arising from the possibility to use this "|" in the rule headers.

5.2.2. CLIENT RULES   These implement something more difficult, which cannot be conveniently implemented in Perl. The body usually consists of a single line with a call of some client program written in C++. An example is the rule

```
VERTEX_BARYCENTER : BOUNDED(true), VERTICES(...)
client("vertex_barycenter", $polyname);
```

which computes the barycenter of the vertices of a bounded polytope from the complete list of vertices in rational coordinates.

5.2.3. INTERFACE RULES   This is the most complex type of rules. The rule body converts data from the **polymake** format to whatever is desired by the other program, then calls the external program and, finally, converts the external program's output into **polymake** format. An example is the rule

```
?porta, VERTICES, N_VERTICES : FACETS | INEQUALITIES, ...
```

which calls PORTA to perform a dual convex hull computation.

5.3. The engine

The **polymake** engine is written in Perl. It reads the rule base, a file containing information about some polyhedron and a user's question about properties of this polyhedron. We do not want to focus on the technical details here. Instead we are concerned with how **polymake** plans the steps which are necessary to derive the requested information about a polyhedron from the given.

During the planning phase a polyhedron, its properties, and the rules can be modelled by means of a weighted directed multigraph in the following way. The description given here is slightly simplified.

The nodes in this graph are combinations of properties (sections) and section dependences. These represent the theoretically possible *states* of a polytope at different times. Each combination of sections stands for the properties which are known. For each of these sections there is a (possibly empty) list of those sections which were used as sources during its production. Thus a node corresponds to a collection of known properties of a polytope plus their history.

Each rule gives rise to several edges of this multigraph indicating that it is possible to move from one state to the other by applying the specific rule without violating certain consistency conditions. More formally, consider a node $\mathfrak{G}$ with a set S of sections and a dependence function $d : S \to 2^S$. The dependences at $\mathfrak{G}$ induce a partial order '$<$' among the sections. We write $s_0 < s_1$ if $s_1 \in d(s_0)$ and likewise for the transitive closure of the relation. Take a rule ρ with a set T_ρ of targets and a set S_ρ of sources. The rule ρ causes an edge from the node $\mathfrak{G} = (S, d)$ to some other node if the following conditions are satisfied.

(i) (Soundness) $S_\rho \cap T_\rho = \emptyset$,
(ii) (Consistency) $\sigma \not< \tau$ for $\sigma \in S_\rho$ and $\tau \in T_\rho$, and
(iii) (Applicability) $S_\rho \subseteq S$.

Now we still have to say which is the target node of the edge starting at $\mathfrak{G}$ induced by ρ. Its set of sections is $S' = S \cup T_\rho \setminus \{s \in S \mid s < t \text{ for some } t \in T_\rho\}$. The new dependence function $d' : S' \to 2^{S'}$ is defined as follows. If $s \in T_\rho$ then $d'(s) = S_\rho$, otherwise $d'(s) = d(s)$.

Each edge has a certain weight (e.g. reflecting the presumed duration of the respective task or some other preference). The multigraph does contain multiple edges between certain pairs of vertices, but it does not contain loops. The polymake server performs a search for a minimum weight path joining the node corresponding to the initial information about the polytope to some node containing the requested information. Essentially, Dijkstra's algorithm is applied, for example see [12, 25.2]. Note that this graph is only implicitly given by the rule base and never explicitly constructed.

The above conditions model the following properties. Soundness says that a rule must not overwrite its sources. Consistency forbids that some source depends on any of the targets. The applicability condition requires that there is sufficient information available to apply the rule. Observe that the first two conditions ask for properties of the rules without reference to a particular state S, while only the third condition refers to the history. Sometimes polymake files contain time stamps as comments (mainly for debugging purposes), but they are not used by the system.

A few more remarks concerning this slightly more formal discussion are appropriate. We have to confess that what we described above is not the whole truth. Instead we discussed a simplified model. Firstly, as already mentioned in 5.2.1, we did not reveal the precise meaning of the "|" which occurred in some of the rules: This basically introduces some notion of a logical or. Secondly, we did not mention the "?" character in some rule headers: On the left hand side they essentially are used as names of the respective rules. This way we can allow the user to decide to use a specific convex hull code, for instance. Thirdly, we did not talk about *attributes* which occur in parentheses after some of the section names. They serve several needs at the same time, which we do not want to discuss here.[3] Finally, strictly speaking, the soundness criterion (i) does not hold for the rule base as of now. This is due to a minor technical detail. Future versions of polymake will have a sound rule base in this sense.

5.4. The generators

polymake comes with a bunch of programs which generate polymake descriptions of popular classes of polytopes. They can be divided into two groups. Generators of the first kind produce a polytope from scratch, while the others need one or more polytopes as input.

The 4-permutahedron whose Schlegel diagram was shown in Figure 8 can be produced as follows.

```
> permutahedron perm4.poly 4
```

---

[3]You can find out a little bit more about attributes from the polymake web documentation, if you really want to. They will vanish in future versions of polymake.

Among others we have generators for simplices, cubes, and cyclic polytopes. It is possible to construct products, convex hulls of unions, intersections, and more.

The following gives an orthogonal projection of the 4-permutahedron onto the subspace spanned by all unit vectors except for the first one.

```
> proj projection.poly perm4.poly - 1
```

There are also generators which apply affine/projective transformations to a given polyhedron. For instance, the following command performs the transformation in Lemma 3.1.

```
> orthantify transformed.poly some_polyhedron.poly
```

Each generator has its individual parameter syntax. But, they all have in common that simply calling the respective program without parameters yields a help message. This is the same as for the clients described below.

```
> proj
usage: proj <outfile> <infile> [-] [<k1> [<k2> ...] ]
```

Each polytope involved in the construction of a new polytope can, of course, be described as the convex hull of points (and rays) or as the intersection of halfspaces. In order to compute, say, the intersection of two polyhedra one can simply merge the facets of both.

```
> intersection i.poly v-polytope.poly h-polyhedron.poly
```

In this example **polymake** might perform a convex hull computation on **v-polytope.poly** before it can merge the facets into one list. Again, this is done automatically.

The generators are written in C++ using the **Poly** class interface to **polymake** and the **polymake** Template Library, cf. Section 5.6.

5.5. The clients

A **polymake** client performs a single task from defined input. For example, there is a client which computes the rank of a matrix and adds an offset. This can be used to compute the dimension of the affine span of a set of points. Like generators, clients can be called form the command line.

```
> rank p.poly POINTS DIM -1
```

In the example the program reads the coefficient matrix from the section **POINTS** in the file **p.poly**, computes its rank, subtracts 1, and finally writes the result into the new section **DIM** of the same file.

Clients overwrite sections if they existed already before the call of the client. Moreover, any section which depends on an overwritten section is erased. This way **polymake** asserts the consistency of all the sections within one file.

Much of what has been said about generators is also true for clients. In particular, a client displays its command line syntax if called without parameters.

```
> rank
usage: rank <file> <matrix_section> <rank_section> [ offset ]
```

It should be mentioned, however, that calling a client from the command line does not happen frequently. More often, these clients are called by the **polymake** engine during the application of the rules. But, there are a few cases, in which it is useful to call a client directly.

For instance, a Schlegel diagram depends on the facet which is projected on as well as the position of the viewpoint. By default, **polymake** always projects onto the first facet (numbered '0'). **polymake** always sets the viewpoint on the line joining the vertex barycenter of the polytope with the vertex barycenter of the projection facet. The user can choose a point on that line by specifying a *zoom factor z* where $0 < z < 1$. Here $z = 1$ means that the viewpoint lies on the facet and $z = 0$ corresponds to infinity. Specifying a value of $z \geq 1$ is allowed, but it does not give a Schlegel diagram. The following example shows how the user can ask **polymake** to project onto the second facet (numbered '1') and at the same time move the viewpoint closer to the facet; z defaults to 0.9.

```
> schlegel perm4.poly 1 0.95
```

The client **schlegel** writes the coordinates for the Schlegel diagram into the section SCHLEGEL_VERTICES. The next time the user asks for a picture of the Schlegel diagram, the previously computed coordinates are used.

```
> polymake perm4.poly SCHLEGEL
```

Clients are also written in C++ using the class library which is described below.

5.6. The polymake template library

The most recent release of the ANSI C++ Standard includes the Standard Template Library (STL), cf. [28]. The **polymake** Template Library (PTL) is built on top of STL and designed in a similar spirit. It features a variety of C++ template classes which are used by the **polymake** clients.

Most importantly, the PTL provides a set of container classes which have been designed with applications in polyhedral geometry/combinatorics in mind. However, they should easily fit various other needs as well. All the container classes have in common that they offer a range of consistency checks which can be enabled at compile-time. If this is switched off, consistency checking does not take place and there is no penalty with respect to efficiency during run-time. Overhead from unnecessary copying of data is almost avoided by using reference counting, cf. Knuth [25].

There are template classes **Vector** and **Matrix** which allow standard Linear Algebra computations such as solving systems of linear equations, inverting matrices, computing scalar products, and so on.

A second group of container classes is devoted to combinatorial objects. There are classes for (ordered) sets, maps (i.e. associative arrays) and graphs. The key difference between the set and map classes from the PTL versus similar STL-classes is that our implementation uses AVL-trees, see Knuth [26].

In addition to the container classes listed above there is a class `Poly` which offers an interface to the **polymake** engine, cf. Section 7.1. This is how *all* the functionality of the whole system becomes available in C++ programs.

5.7. Arithmetic

By default all relevant computations with coordinates are performed with arbitrary precision. **polymake** uses the GNU Multiprecision Library. For convenience we developed C++-wrappers to this C-library which can also be used independently of the rest of the **polymake** Template Library. There are two classes, `Integer` and `Rational` with the expected semantics. The interfaces are (as far as possible) identical to the old `libg++` classes with the same names.

It is also possible to use floating point arithmetic instead by specifying `-f float.rules` as an additional parameter in the command line call. This can be desirable in two cases. Either to speed up computations, or to work with polytopes which do not have a rational representation. However, the usefulness is limited, because practically no provisions exist to assert numerical stability of the results. For convex hull computations with floating point arithmetic **polymake** interfaces to CDD. It does happen that a combinatorial description of a polytope obtained this way is false in the sense that there is *no* polytope with the claimed incidences.

6. The polytope database

The polytope database is intended to complement the set of polytopes which belong to some suitably parameterized family. While the latter ones can be produced by generators, cf. Section 5.4, the database contains individual polytopes which are interesting for various reasons.

The polytope database is available on the Internet from the **polymake** home page at URL **http://www.math.tu-berlin.de/diskregeom/polymake/**. Contributions are most welcome.

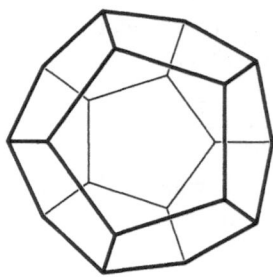 

FIGURE 13. The two exceptional regular 3-polytopes.

6.1. Regular polytopes

Without any doubt, the most classical polyhedra are the five Platonic Solids: the tetrahedron, the cube and the octahedron, the icosahedron and the dodecahedron. They are the only 3-dimensional *regular* polytopes, i.e. they have pairwise congruent facets as well as pairwise congruent vertex figures. They are closely related to finite reflection groups, cf. Coxeter [13], Humphreys [21].

There are three infinite families of regular d-polytopes: the simplices, the cubes and the cross-polytopes. For each of these classes there is a corresponding `polymake` generator: `simplex`, `cube`, `cross`. It turns out that there are only 5 regular polytopes which do not belong to any of these families: these are the icosahedron and the dodecahedron in dimension 3 (cf. Figure 13), and the 24-cell, the 600-cell and the 120-cell in dimension 4. These five exceptional regular polytopes are contained in the polytope database. Note that, among the exceptional regular polytopes, only the 24-cell has a rational realization (as a regular polytope), cf. Figure 14.

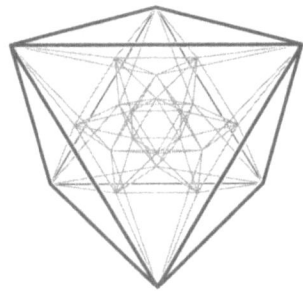

FIGURE 14. Schlegel diagram of the 24-cell.

6.2. 0/1-polytopes

Polytopes which have vertices with 0/1-coordinates only play an important role in combinatorial optimization. The polytope library contains some particularly interesting examples, such as the (up to cube symmetry) unique 5-dimensional 0/1-polytope with 40 facets. By the classification of 5-dimensional 0/1-polytopes according to Aichholzer [1] this is maximal.

```
VERTICES
1  0 0 0 0 0
1  1 0 1 0 0
1  1 1 1 0 0
1  0 1 0 1 0
1  0 1 1 1 0
1  0 0 1 0 1
1  0 0 1 1 1
```

```
1  1 0 0 1 0
1  1 0 0 1 1
1  0 1 0 0 1
1  1 1 0 0 1
1  1 1 1 1 1
```

All 0/1-polytopes of dimension at most 5 are available in **polymake** format thanks to Aichholzer's implementation, cf. [1].

7. Extending the functionality

The **polymake** system is highly customizable and extendible. However, it requires a bit of programming to extend the system.

7.1. The C/C++ interface

All **polymake** generators and clients are written in C++, cf. Stroustrup [31]. They use an interface to the **polymake** system which is implemented in a special class **Poly** derived from the usual **iostream** class. Since version 1.3 the code is designed for compilation with the GNU C++ compiler egcs 1.1.

From the software point of view, clients are more general than generators (because they read and write, while generators only write). Therefore we explain the rest in terms of clients only.

The **Poly** class takes care of the data exchange between the client and the **polymake** server. Each execution of a client is a transaction which can be subdivided into several read and write phases by the client. If the client announces to read a section, then the **polymake** server tries to provide the data. The server applies several rules if necessary, thereby possibly calling other clients. Writing is less subtle. Successfully written data must be acknowledged by the client.

This is the code of the client **rank** already discussed above, cf. Section 5.5. We stripped some **#include** statements plus the parameter checks.

```
int main (int argc, char *argv[]) {

Poly p (argv[1],ios::in+ios::out);

p.read(argv[2], 0);
p.write(argv[3], 0);

Matrix<Rational> M(p);
p << rank(M)+(argc==5 ? atoi(argv[4]) : 0) << "\n" << endl;

return 0;
}
```

The code uses the template class **Matrix** which is part of the PTL, cf. Section 5.6. The function **rank()** is a friend function of this class. The **Rational** class is a

C++-wrapper class to the GNU Multiprecision Library. The functions read() and write() are members of the Poly class. They announce reading and writing of a list of sections (terminated by '0'), respectively. The operators >> and << perform the read and write operations, respectively. Successful writing is acknowledged by sending an empty line.

On the technical level the communication is based on sockets. But we do not want to go into the details here. There is also a similar C-interface to polymake.

The key advantage in using C++ instead of C relies on the PTL. It provides data structures such as vectors, matrices and graphs which allow for programming on a rather abstract level.

7.2. Extending the rule base

Each rule has a header which describes the functionality of the rule. The rule body implements this functionality; it is written in Perl, cf. Wall, Christiansen, Schwartz [32].

Essentially the rule body is the body of a Perl subroutine which signals successful execution by returning a non-zero value. polymake guarantees that a rule body behaves as it is announced in the rule header. In particular, a rule body is only allowed to read from its sources and to write to its targets. Special subroutines give() and take() take care of reading from and writing to sections.

Some of the rules just call a client. This is the rule which computes the dimension of a polyhedron. It tries to read the VERTICES section or, if this is not available without further computation, the POINTS section. In either case the rank of the respective coordinate matrix is determined by the client rank.

```
DIM : VERTICES | POINTS
client("rank", $polyname, "VERTICES | POINTS", "DIM", "-1")
```

8. Future plans

The polymake system is under continuous development. This includes the augmentation of polymake's mathematical knowledge as well as improvements on the technical level. Let us sketch some of the things we have in mind.

In the near future polymake will learn a little bit about oriented matroids and triangulations by interfacing to Rambau's TOPCOM [29]. This allows for exact volume computation and plenty more.

One of the next versions of polymake will provide the user with a shell-type environment, similar to most Computer Algebra Systems. It will be possible to work with polytopes (and other things) in an object-oriented manner. polymake will be scriptable with Perl as a full-featured programming language. polymake's file format will be changed from an ASCII based format to a binary format; of course, backward compatibility is assured.

In the (probably far) future there should be some sort of distributed polytope database on the Internet, which allows to store huge classes of polytopes. We think that such a project should be worth the joint effort of several groups.

Acknowledgments. The `polymake` system was born in 1996. Since then `polymake` has continuously been extended and improved. Many people contributed in various ways. Let us mention a few of them.

Among the first users of the system were Kerstin Fritzsche, Andrea Höppner, and Wolfram Schlickenrieder. They had to struggle through the many bugs, design flaws, etc. of the early stage. Carsten Jackisch and Gerald Stein contributed some of the client code. Alexander Schwartz helped with the HTML documentation. Volker Kaibel and Marc Pfetsch gave helpful comments on a previous version of this paper. Günter M. Ziegler initiated this project. `polymake` could not have evolved without his constructive criticism and his ongoing support. The German-Israeli-Foundation partially funded `polymake`, grant I-0309-146.06/93.

References

[1] O. Aichholzer, *01poly*,
http://www.cis.tu-graz.ac.at/igi/oaich/info01poly.html, 1999.

[2] N. Amenta, *Computational geometry software*, In Goodman and O'Rourke [19], pp. 951–960.

[3] D. Avis, *lrs: A revised implementation of the reverse search vertex enumeration algorithm*, this volume, 177–198.

[4] ———, *lrs, Version 3.2*, ftp://mutt.cs.mcgill.ca/pub/C/lrs.html, Oct 14, 1998.

[5] D. Avis, D. Bremner, and R. Seidel, *How good are convex hull algorithms?*, Comput. Geom. **7** (1997), no. 5-6, 265–301.

[6] D. Avis and K. Fukuda, *A pivoting algorithm for convex hulls and vertex enumeration of arrangements and polyhedra.*, Discrete Comput. Geom. **8** (1992), no. 3, 295–313.

[7] M.M. Bayer and C.W. Lee, *Combinatorial aspects of convex polytopes*, pp. 485–534, In Gruber and Wills [20], 1993.

[8] G. Blind and R. Blind, *Convex polytopes without triangular faces*, Israel J. Math. **71** (1990), no. 2, 129–134.

[9] ———, *Cubical 4-polytopes with few vertices*, Geom. Dedicata **66** (1997), no. 2, 223–231.

[10] D. Bremner, *Polytope Base, Version 0.2*,
http://www.math.washington.edu/~bremner/PolytopeBase/, Mar 4, 1999.

[11] T. Christof and A. Loebel, *PORTA: POlyhedron Representation Transformation Algorithm, Version 1.3.2*, http://www.iwr.uni-heidelberg.de/iwr/comopt/soft/PORTA/readme.html, Nov 24, 1998.

[12] T.H. Cormen, C.E. Leiserson, and R.L. Rivest, *Introduction to algorithms*, MIT Press, 1990.

[13] H.S.M. Coxeter, *Regular polytopes*, Dover, 1973.

[14] H. Edelsbrunner, *Algorithms in combinatorial geometry*, Springer, 1987.

[15] C. Young et al., *POV-Ray: Persistence Of Vision, Version 3.1,*
http://www.povray.org/, 1999.

[16] F.J. Brandenburg et al., *Graphlet, Version 5.0,* http://www.fmi.uni-passau.de/
Graphlet/, Jan 11, 1999.

[17] K. Fukuda, *cddplus, Version 0.76a,*
http://www.ifor.math.ethz.ch/staff/fukuda/cdd_home/cdd.html, Jun 8, 1999.

[18] K. Fukuda and A. Prodon, *Double description method revisited,* LNCS, vol. 1120,
1996.

[19] J.E. Goodman and J. O'Rourke (eds.), *Handbook of discrete and computational geometry,* CRC Press, 1997.

[20] P.M. Gruber and J.M. Wills (eds.), *Handbook of convex geometry,* North-Holland,
1993.

[21] J.E. Humphreys, *Reflection groups and Coxeter groups,* Cambridge Univ. Press,
1992, corrected paper-back ed.

[22] M. Joswig, *Reconstructing a non-simple polytope from its graph,* this Volume, 167–
176.

[23] M. Joswig and G.M. Ziegler, *Neighborly cubical polytopes,* Discrete Comput. Geometry (to appear), math.CO/9812033.

[24] G. Kalai, *Linear programming, the simplex algorithm and simple polytopes,* Math.
Program. Ser. B **79** (1997), 217–233.

[25] D.E. Knuth, *The art of computer programming, I. Fundamental algorithms,* 3rd ed.,
Addison-Wesley, 1997.

[26] _____ , *The art of computer programming, III. Sorting and searching,* 2nd ed.,
Addison-Wesley, 1997.

[27] S. Levy, T. Münzner, and M. Phillips, *Geomview, Version 1.6.1,*
http://www.geom.umn.edu/software/geomview/, Oct 30, 1997.

[28] D.R. Musser and A. Saini, *STL tutorial and reference guide,* Addison-Wesley, 1996.

[29] J. Rambau, *TOPCOM, Version 0.2.0,*
http://www.zib.de/rambau/topcom.html, Jul 28, 1999.

[30] J. Richter-Gebert and U.H. Kortenkamp, *The interactive geometry software Cinderella,* Springer, 1999.

[31] B. Stroustrup, *The C++ programming language,* Addison-Wesley, 1997, 3rd ed.

[32] L. Wall, T. Christiansen, and R.L. Schwartz, *Programming Perl,* O'Reilly, 1996, 2nd
ed.

[33] G.M. Ziegler, *Lectures on polytopes,* Springer, 1998, 2nd ed.

Ewgenij Gawrilow, Michael Joswig
Fachbereich Mathematik, MA 7–1
Technische Universität Berlin
Straße des 17. Juni 136
10623 Berlin, Germany
[gawrilow,joswig]@math.tu-berlin.de

Flag Numbers and FLAGTOOL

Gil Kalai, Peter Kleinschmidt and Günter Meisinger

Abstract. FLAGTOOL is a computer program for proving automatically theorems about the combinatorial structure of polytopes of dimensions at most 10. Its starting point is the known linear relations (equalities and inequalities) for flag number of polytopes. After describing the state of the art concerning such linear relations we describe various applications of FLAGTOOL and we conclude by indicating several directions for future research and automation. As an appendix we describe FLAGTOOL's main tools and demonstrate one working session with the program.

1. Face numbers, flag numbers, g-numbers and convolutions

1.1. Face numbers

For a d-polytope P the number of its k-faces is denoted by $f_k(P)$. The vector

$$(f_0(P), f_1(P), \ldots f_{d-1}(P))$$

is called the f-vector of P.

A basic problem (perhaps hopeless for $d > 3$) is:

Problem 1.1. *Characterize all the f-vectors of d-polytopes.*

A more realistic subproblem is:

Problem 1.2. *Find all the linear relations (linear equalities and linear inequalities) among face numbers of d-polytopes*

It is well known [Gr67, Zie95] that the only linear equality that holds among face numbers of d-polytopes is Euler's relation:

$$\sum_{i=0}^{d-1} f_i(P) = 1 + (-1)^d.$$

As for inequalities for $d > 5$ the only known linear inequalities are quite trivial:

$$f_r(P) \geq \binom{d+1}{r+1}, \quad 2f_1(P) \geq df_0(P), \quad 2f_{d-2}(P) \geq df_{d-1}(P).$$

1.2. Flag numbers

For a d-polytope P, and a subset $S = \{i_1, \ldots, i_k\} \subset \{0, 1, \ldots, d-1\}$ the flag number $f_S^d(P)$ is the number of chains of faces of P $F_1 \subset F_2 \subset \cdots \subset F_k$ such that $\dim F_j = i_j$. We will omit the superscript d if its value is clear from the context. (The same definition applies to ranked lattices.) The vector of flag numbers $f_S(P)$ (where the indices are ordered according to some fixed ordering) is called the flag vector of P. For simplicial polytopes the flag numbers are determined by the face numbers, but for general polytopes flag numbers seems to be the "correct" invariants.

Problem 1.3. *Characterize flag vectors of d-polytopes.*

Again this may be hopeless and a more realistic task is:

Problem 1.4. *Find all the linear relations among flag numbers of d-polytopes*

We will denote by $\mathcal{A}_d$ the affine space spanned by flag vectors of d-polytopes and by $\mathcal{P}_d \subset \mathcal{A}_d$ the cone spanned by flag vectors of d-polytopes.

Remark: While these problems on flag numbers are more general than the corresponding problems for face-numbers, deriving conclusions for the face numbers from information on flag numbers may also be a non-trivial task.

1.3. The theorem of Bayer and Billera

A remarkable theorem of Bayer and Billera [BayB85] asserts that the affine dimension of the space $\mathcal{A}_d$ of flag vectors of d-polytopes is $c_d - 1$, were c_d is the d-th Fibonacci number.

Bayer and Billera used Euler's formula to deduce the following relation commonly referred to as the "generalized Dehn-Sommerville relations":

Let $S \subset \{0, 1, \ldots d-1\}$, $k \in S \cup \{-1, d\}$ and $i \leq k-2$. If S contains no integer between i and k then

$$\sum_{j=i+1}^{k-1} f_{S \cup \{j\}}^d(P) = (1 - (-1)^{k-i-1}) f_S^d(P).$$

Bayer and Billera also showed that these relations span all the affine relations among flag numbers of d-polytopes.

1.4. Bases for flag vectors and bases for polytopes

A d-form will denote a linear combination of flag numbers f_S^d. A *basis* for the space of flag numbers of d-polytopes is a collection of d-forms which affinely span the space $\mathcal{A}_d$. The *special* flag numbers are those flag numbers f_S^d such that $S \subset \{1, 2, \ldots d-2\}$ and S does not contain two consecutive integers. It follows from the generalized Dehn-Sommerville relation that every flag number can be represented as an affine combination of special flag numbers. Other bases of flag numbers of d-polytopes were found in [Kal88, BilL].

A *basis* of polytopes is a collection of c_d polytopes whose flag vectors are affinely independent. See [BayB85, Kal88] for two such constructions.

1.5. h- and g-numbers for simplicial polytopes

Let $d > 0$ be a fixed integer. Given a sequence $f = (f_0, f_1, \ldots, f_{d-1})$ of nonnegative integers, put $f_{-1} = 1$ and define $h[f] = (h_0, h_1, \ldots, h_d)$ by the relation

$$\sum_{k=0}^{d} h_k x^{d-k} = \sum_{k=0}^{d} f_{k-1}(x-1)^{d-k}. \tag{1}$$

If $f = f(P)$ is the f-vector of a simplicial d-polytope P then $h[f] = h(P)$ is called the h-vector of P. The g-vector $g(K) = (g_0, g_1, \ldots, g_{[d/2]})$ of P is defined by $g_i = h_i - h_{i-1}$. Thus, $g_0 = 1$, $g_1 = f_0 - (d+1)$, $g_2 = f_1 - df_0 + \binom{d+1}{2}$ and $g_3 = f_2 - (d-1)f_1 + \binom{d}{2}f_0 + \binom{d+1}{3}$ and so on.

In 1970 McMullen [McM71] proposed a complete characterization of f-vectors of boundary complexes of simplicial d-dimensional polytopes. McMullen's conjecture was settled in 1980. Billera and Lee [BiLe81] proved the sufficiency part of the conjecture and Stanley [Sta80] proved the necessity part. Stanley's proof relies on deep algebraic machinery including the hard Lefschetz theorem for toric varieties. Recently, McMullen [McM93] found a self-contained proof of the necessity part of the g-theorem. It is conjectured that the g-theorem applies to arbitrary simplicial spheres.

For positive integers $n \geq k > 0$ there is a unique expression of n of the form

$$n = \binom{a_k}{k} + \binom{a_{k-1}}{k-1} + \cdots + \binom{a_i}{i}, \tag{2}$$

where $a_k > a_{k-1} > \cdots > a_i \geq i > 0$. This given, define

$$\partial^k(n) = \binom{a_{k-1}-1}{k-1} + \binom{a_{k-1}-1}{k-2} + \cdots + \binom{a_i-1}{i-1}. \tag{3}$$

Theorem 1.1 (The g-theorem). *For a vector $h = (h_0, h_1, \ldots, h_d)$ of nonnegative integers the following conditions are equivalent:*

(i) h *is the h-vector of some simplicial d-polytope.*
(ii) h *satisfies the following conditions*
    (a) $h_k = h_{d-k}$ *for* $k = 0, 1, \ldots, [\frac{d}{2}]$
       Put $g_k = h_k - h_{k-1}$.
    (b) $g_0 = 1$ *and* $g_k \geq 0$, $k = 1, 2, \ldots, [\frac{d}{2}]$.
    (c) $\partial^k(g_{k+1}) \leq g_k, k < [\frac{d}{2}]$

1.6. h- and g-numbers for general polytopes

Intersection homology theory has led to deep and mysterious extensions of h- and g-numbers from simplicial polytopes to general polytopes. The definition [Sta94''] goes as follows. For a polytope P denote by P_k the set of k-faces of P. Define by induction two polynomials

$$h_P(x) = \sum_{k=0}^{d} h_k^d x^{d-k}, \quad g_P(x) = \sum_{k=0}^{[d/2]} g_k^d x^{d-k},$$

by the rules: (a) $g_k^d = h_k^d - h_{k-1}^d$, (b) If P is the empty polytope or a 0-polytope P, $h_P = g_P = 1$, and

$$h_P(x) = \sum_{k=0}^{d} (x-1)^{d-k} \sum \{g_F(x) : x \in P_k\}.$$

Thus $g_1^d(P) = f_0(P) - d - 1$ and

$$g_2^d(P) = f_1(P) + \sum \{f_0(F) - 3 : F \in P_2\} - df_0(P) + \binom{d+1}{2}.$$

The value of g_2^d for general polytopes has also a rigidity theoretic meaning and is nonnegative for every polytope. The nonnegativity of g_2^d is still open for more general objects like polyhedral spheres and manifolds. It follows from intersection homology theory for toric varieties that g_k^d is nonnegative for every rational polytope. This is still open for general polytopes.

Problem 1.5. *Characterize g-vectors of d-polytopes.*

It is conjectured that g-vectors of arbitrary d-polytopes satisfy (and therefore are characterized by) the same non linear relations which were proved for simplicial polytopes.

Problem 1.6. *What is the significance of the g-numbers for the combinatorial theory of d-polytopes?*

1.7. Duality of polytopes

For a d-polytope P we denote by P^* the dual of P. There is an order-reversing bijection between faces of P and faces of P^*. Put $\bar{g}_k^d(P) = g_k^d(P^*)$. Clearly $\bar{g}_k^d(P)$ is nonnegative for every rational polytope P.

There are various connections between flag numbers of polytopes and their dual which are quite mysterious and are related to mirror symmetry. See [Sta92, BaBo96, Kal88].

1.8. Intervals in face-lattices of polytopes

The set of faces of a d-polytope form a ranked poset of rank $d+1$ with the lattice property. We will use the dimension as the grading and thus the empty face will have grade -1. Intervals in the face lattices of polytopes are themselves face lattices of polytopes, see [Gr67, Zie95, Kal88]. Intervals of type $[a, b]$ are intervals $[F, G]$ where $\dim F = a$ and $\dim G = b$. If $a = -1$ then F is the empty face and $[F, G]$ is simply the face lattice of G.

1.9. Convolutions

Let m^d, m^e be linear combinations of flag numbers of d- and e-polytopes respectively. For a polytope P of dimension $d + e + 1$ define the convolution of m^d and m^e by

$$m^d * m^e(P) = \sum \{m^d(F) \cdot m^e(P/F) : F \text{ a } d\text{-face of } P\}.$$

For a Hopf-algebraic treatment of flag numbers and their convolutions, see [BilL]. The following lemma [Kal88] is immediate.

Lemma 1.2.

(1) $m^d * m^e(P)$ *is a linear combination of flag numbers of* $(d + e + 1)$*-polytopes.*

(2) *If* $m^d(P) = 0$ *for every* d*-polytope* P *or* $m^e(Q) = 0$ *for every* e*-polytope* Q *then* $m^d * m^e(R) = 0$ *for every* $d + e + 1$*-polytope* R*.*

(3) *If* $m^d(P) \geq 0$ *for every* d*-polytope* P *and* $m^e(Q) \geq 0$ *for every* e*-polytope* Q *then* $m^d * m^e(R) \geq 0$ *for every* $d + e + 1$*-polytope* R*.*

Convolutions of the g_i's and $\bar{g}_i$'s yield a large collection of linear inequalities for flag numbers of d-polytopes. We will denote by $\mathcal{Q}_d$ the cone of flag numbers described by all these inequalities. As it turns out the simple inequalities $f_i^d(P) \geq \binom{d+1}{i+1}$ do not follow from convolutions of the g_i's. We denote by $\mathcal{Q}_d'$ the cone of flag numbers obtained by adding these inequalities, their polars and the derived inequalities by convolutions.

1.10. The cd-index

Remarkable classes of invariants for Eulerian posets are given by Fine's cd-index. See [BayK91, Sta94']. The cd-index for d-dimensional polytopes is a polynomial of degree d in two non-commuting variables of degrees 1 and 2, respectively. This polynomial has c_d coefficients, each one of which is a d-form. (Namely, a linear combination of flag numbers of d-polytopes) and together they consist of a basis of such forms. It was proved by Stanley [Sta94'] that these coefficients are nonnegative for every d-polytope, and Billera and Ehrenborg [BiEh] proved that the values of these forms are at least as large as their value for the d-simplex. In low dimensions these inequalities already follows from those in $\mathcal{Q}_d'$.

2. FLAGTOOL

The previous section shows that the Generalized Dehn Sommerville Equations and the nonnegativity of convolutions of the numbers g_i^k and $\bar{g}_i^k$ ($0 \leq k \leq d$, $0 \leq i \leq \lfloor k/2 \rfloor$) yield a lot of linear relations between the flag numbers of general d-polytopes for a fixed dimension d. It is very hard to compute them or to derive new results from them without using a computer because the number of those relations is large already for small dimensions. Therefore, Meisinger [Mei94] developed a program called FLAGTOOL. The main purpose of this program is to

- compute all (known) linear relations between the flag numbers of general d-polytopes for small dimensions, say $3 \leq d \leq 10$,
- extract and automatically prove new results from those relations.

The aim of the following section is to present the main features of the program.

2.1. Basic ideas and motivation

In 1990 Kalai [Kal90] proved that every d-polytope ($d \geq 5$) has a 2-face with less than 5 vertices. This implies that there does not exist a 5-polytope all 2-faces of which are pentagons. The proof was obtained by taking for dimension five all known linear inequalities for flag numbers and all possible convolutions of the inequality corresponding to the negation of the theorem's claim ($f_0^2 - 5 \geq 0$ for intervals of type $[-1, 2]$). The resulting set of linear inequalities, as an input of a Linear Programming Problem, had no feasible solution and therefore the correctness of the theorem followed.

This suggests that more results can be proved in a similar way by using this idea systematically. The main aspect for the development of FLAGTOOL is based on this idea and can be summarized as follows. We include examples for a better understanding. See also Figure 1 for the general scheme of supporting theorem proving with FLAGTOOL.

1. Consider a fixed dimension d. FLAGTOOL works with dimensions $3 \leq d \leq 10$. This value can be increased if it does not exceed the computer's memory capacity or causes runtime problems.

2. For a fixed dimension d we derive all known nonnegative d-forms from the numbers g_i^k and $\overline{g_i}^k$ ($0 \leq k \leq d$, $0 \leq i \leq \lfloor k/2 \rfloor$) and their convolutions. (It turned out that a lot of the resulting inequalities are redundant. FLAG-TOOL omits those redundant inequalities and computes a system B of linear inequalities in terms of flag numbers for each dimension d.)

3. (a) In order to prove a linear inequality for flag or face numbers in this fixed dimension d, we have to add the negation of this inequality to the system B. To prove $-2f_1 + 3f_2 - 2f_3 \geq 0$ which holds for all 5-polytopes, we add the negation $2f_1 - 3f_2 + 2f_3 - 1 \geq 0$.

   (b) To prove facts about low dimensional faces or quotients, we have to add inequalities (negation of conjectured inequalities) to intervals (in the face-lattices) of a d-polytope and to convolve them with the numbers g_i^k and $\overline{g_i}^k$ ($0 \leq k \leq d$, $0 \leq i \leq \lfloor k/2 \rfloor$) to create a system of inequalities which contains the system B. To prove the result that 5-polytopes always have a 2-face with less than 5 vertices, we have to add the 2-dimensional inequality $f_0^2 - 5 \geq 0$ to the bottom interval $[-1, 2]$.

4. Express all the new linear inequalities in terms of special flag numbers. This results in a set A (that contains B) of linear inequalities.

5. If the set A of linear inequalities has no feasible solution, the negation of at least one of the added inequalities is true for all rational polytopes. If only g_0^d, g_1^d and g_2^d are used in convolutions, the result holds for all polytopes. The infeasibility can be proved for example by using phase I of an LP-solver or symbolic mathematical programs.

The program itself consists of a set of subtools which are described briefly in the Appendix. For more details see the user manual of FLAGTOOL [Mei94].

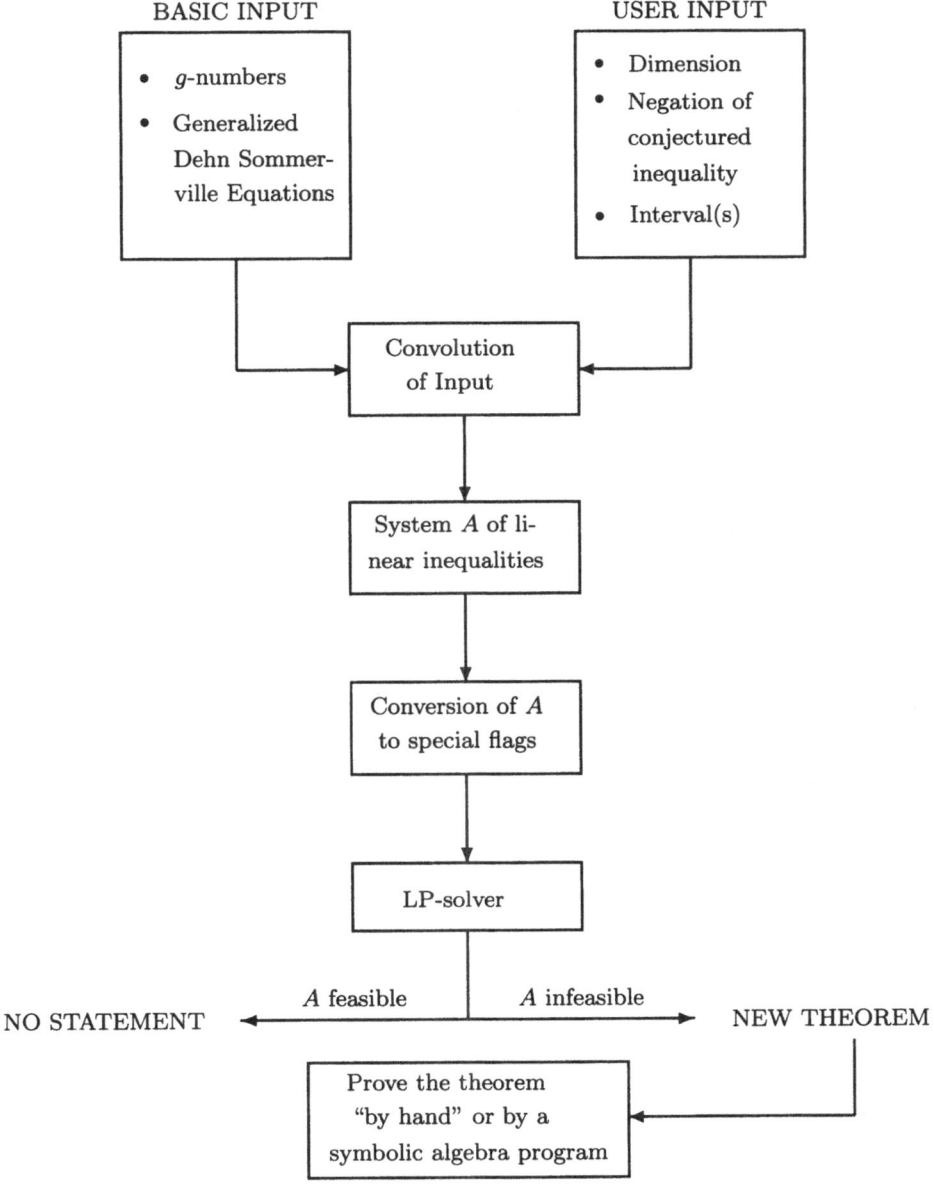

FIGURE 1. Automated theorem proving by FLAGTOOL

3. The linear cone of flag vectors and f-vectors

3.1. What are all the linear inequalities for flag numbers?

It was conjectured in [Kal88] that $\mathcal{P}_d = \mathcal{Q}_d$, i. e., that every linear inequality on the flag numbers of general d-polytopes is equivalent to the nonnegativity of some nonnegative combination of convolutions of the g_i^k and $\overline{g}_i^k$ $(0 \leq k \leq d,$ $0 \leq i \leq \lfloor k/2 \rfloor)$. It turned out that this conjecture is false.

Proposition 3.1. *The linear inequality* (*) $f_2 - 35 \geq 0$, *which holds for all 6-polytopes, is not a nonnegative combination of convolutions of the numbers g_i^k and of $\overline{g}_i^k$ $(0 \leq k \leq 6, 0 \leq i \leq \lfloor k/2 \rfloor)$.*

Proof. We determined using FLAGTOOL all the linear inequalities for the cone $\mathcal{Q}_6$. The point

$$(f_0, f_1, f_2, f_3, f_4, f_{02}, f_{03}, f_{04}, f_{13}, f_{14}, f_{24}, f_{024})$$
$$= (7, 21, 0, 0, 21, 105, 350, 315, 630, 840, 630, 2520)$$

satisfies all these inequalities but violates (*).   □

3.2. Flag numbers of 4-polytopes

Euler's theorem easily implies a complete description of face numbers (hence flag numbers) of 3-polytopes. In the Appendix we give a list of the known nonredundant inequalities for 4-, 5- and (rational) 6-polytopes.

The situation for 4-dimensional polytopes was considered by Barnette [Ba74] and Bayer [Bay87]. Bayer described the cone $\mathcal{Q}_4$ and in particular identified its seven extreme rays. It is not hard to show that four of these rays are indeed extreme rays of $\mathcal{P}_4$ the cone of flag numbers of 4-polytopes. In order to show that $\mathcal{Q}_4$ describes all linear inequalities among 4-polytopes certain constructions of 4-polytopes are needed. For examples an infinite class of self-dual 2-simplicial polytopes with vanishing g_2 will take care of one such ray. An infinite family of self-dual 2-simplicial polytopes so that the ratio $\frac{f_2}{f_1}$ is unbounded will take care of a second ray.

On the other hand, Billera and Ehrenborg conjectured that for 4-polytopes

$$10 f_0 - f_1 + 9 f_3 - 2 f_{03} \geq 45.$$

They showed that if true this relation added to $\mathcal{Q}_4$ would characterize the linear cone of flag numbers of 4-polytopes.

Note the interesting consequences of this inequality for 2-simplicial 2-simple 4-polytopes. For all such polytopes $f_0 = f_3$ and $f_1 = f_2$ and $f_{03} = f_0 + 2f_1$. The inequality of Billera and Ehrenborg would be for such polytopes $f_1 \leq 9f_0 - 45$, namely it will give a linear upper bound for the number of edges in terms of the number of vertices.

3.3. What are all the linear inequalities for face numbers?

When studying the linear inequalities for the flag numbers of arbitrary d-polytopes the question arises what information on the ordinary f-vector can be derived from those inequalities. For a fixed dimension $d \geq 4$ we have a set of n linear inequalities $A_1 \geq 0, \ldots, A_n \geq 0$ obtained by convolution of the numbers g_i^k and $\overline{g_i}^k$ ($0 \leq k \leq d$, $0 \leq i \leq \lfloor k/2 \rfloor$). Each A_j ($1 \leq j \leq n$) is a d-form in $c_d - 1$ variables (the special flag numbers without $f_\emptyset$) and $c_d - d$ variables among them are not face numbers.

Information about the ordinary f-vector can be obtained by projecting the cone of flag-vectors onto the space of f-vector. We obtained the image of $\mathcal{Q}'_d$ under this projection by a successive elimination of the $c_d - d$ variables which are not face numbers. Elimination of a variable f_S (f_S a special flag number) here means to generate all possible nonnegative combinations of the A_i where f_S does no longer appear.

For dimensions 4 and 5 projecting $\mathcal{Q}'_d$ into the space of face-numbers gave, beside the inequalities mentioned in the introduction, two nontrivial inequalities which were found earlier by Bayer and Kalai. Bayer [Bay87] showed that for 4-polytopes

$$g_2^4 + g_0^0 * g_1^2 * g_0^0$$
$$= (f_{02} - 3f_2 + f_1 - 4f_0 + 10) + (6f_1 - 6f_0 - f_{02})$$
$$= -3f_2 + 7f_1 - 10f_0 + 10 \geq 0.$$

and Kalai [Kal88] showed that for 5-polytopes

$$g_1^2 * g_1^2 + g_0^0 * g_1^2 * g_0^0 + g_0^0 * g_1^1 * g_0^0$$
$$= (-6f_3 + 3f_{03} - f_{13} - 3f_{02} + 9f_2)$$
$$+ (2f_{13} - 3f_{03})$$
$$+ (-6f_1 - f_{03} - f_{13} + 3f_{02})$$
$$= 6f_1 - 9f_2 + 6f_3 \geq 0$$
$$\Rightarrow 2f_1 - 3f_2 + 2f_3 \geq 0$$

No further inequalities for 4- and 5-polytopes are obtained by projecting $\mathcal{Q}'_4$ and $\mathcal{Q}'_5$.

In dimension $d = 6$ we have 28 nonredundant linear inequalities $A_1 \geq 0, \ldots, A_{28} \geq 0$ obtained by FLAGTOOL. When we eliminated successively the variables f_{02}, f_{03}, f_{04}, f_{13}, f_{14} and f_{024} and remove redundancy the result is that no new linear inequality for the ordinary f-vector is obtained.

If no further inequalities for face numbers exist this would imply, for example, that there is a sequence P_n of 6-polytopes so that for every $k \neq 3$, $f_3(P) = o(f_k(P_n))$ as n tends to infinity. This seems very unlikely. Barany conjectured that for every d-polytope $f_k(P) \geq min\{f_0(P), f_{d-1}(P)\}$ and this seems very likely albeit beyond our reach.

4. Low dimensional faces and quotients of high dimensional polytopes

4.1. Some basic conjectures

Much of the rest of the paper is related to the following three conjectures:

Conjecture 4.1. *For every integer $k > 0$ there exist integers $n(k)$ and $d(k)$ so that every d-polytope $d \geq d(k)$ has a k-dimensional face with at most $n(k)$ vertices.*

It can be conjectured that $n(k)$ can be chosen to be 2^k and that the following stronger conjecture holds

Conjecture 4.2. *For every integer $k > 0$ there exists $d'(k)$ so that every d-polytope, $d \geq d'(k)$ has a k-dimensional face which is either a simplex or combinatorially isomorphic to a cube.*

Conjecture 4.3 (Perles). *For every integer $k > 0$ there exists $f(k)$ so that every d-polytope $d \geq f(k)$ has a k-dimensional quotient which is a simplex.*

For simple polytopes the first conjecture follows from a fundamental result of Nikulin, see [Nik86, Kal90]. The second conjecture is open even for simple polytopes.

And finally,

Problem 4.4. *For which integers k and d is it true that for every d-polytope, either P or its polar P^* has a k-dimensional face which is a simplex.*

4.2. Small low dimensional faces of high dimensional polytopes

Theorem 4.1. *Every rational d-polytope $(d \geq 9)$ has a 3-face with less than 78 vertices or 78 facets. Using the simple relation $f_0 \leq 2f_2 - 4$, which holds for all 3-polytopes, this implies that there exists always a 3-face with less than or equal to 150 vertices.*

Proof. Assume that there exists a 9-polytope in which every 3-face has at least 78 vertices and at least 78 facets. This assumption can be expressed by the inequalities $f_0^3 - 78 \geq 0$ and $f_2^3 - 78 \geq 0$. A system of 53 linear 9-forms obtained by convolutions of the g-numbers and of these two added inequalities (in the bottom interval $[-1, 3]$) has no nonnegative feasible solution and therefore Theorem 4.1 is proved. See [MKK]. $\square$

The proof was obtained as follows. FLAGTOOL creates 227 linear inequalities which contain the 53 inequalities above. The infeasibility was first checked using phase I of the LP-solver CPLEX and then proved using the symbolic mathematical program MAPLE V. It seems impossible to prove the infeasibility "by hand." The details of this proof (which for a computer generated proof is quite short) do not seem to contribute to our (human) insight for understanding why the theorem is true.

4.3. Small $(k+1)$-faces for high dimensional k-simplicial polytopes

Conjecture 4.2 would imply that for high enough dimension d every 2-simplicial polytope contain a k-dimensional face which is a simplex. In this section we show that every k-simplicial $(2 \leq k \leq 3)$ d-polytope P (all k-dimensional faces of P are simplices) has small $(k+1)$-dimensional faces if the dimension d is high enough.

Theorem 4.2. *Every 2-simplicial d-polytope $(d \geq 5)$ has a 3-face with less than 8 vertices.*

Proof. It suffices to show that Theorem 4.2 holds for 5-polytopes, because every 5-dimensional face of every 2-simplicial d-polytope $(d > 5)$ is again 2-simplicial and therefore has a small 3-face.

Assume that every 3-face of a 2-simplicial 5-polytope has 8 or more vertices. This assumption is expressed by the inequality $f_0^3 - 8 \geq 0$ in the bottom interval $[-1, 3]$. Consider the following five inequalities for 5-polytopes obtained by convolutions of the g-numbers and the added inequality.

[1] $\qquad g_0^1 * g_1^2 * g_0^0 = -6f_1 - f_{13} + 3f_{02} \geq 0$

[2] $\qquad (f_0 - 8) * g_0^1 = f_{03} - 8f_3 \geq 0$

[3] $\qquad g_1^2 * g_1^2 = -6f_3 + 3f_{03} - f_{13} - 3f_{02} + 9f_2 \geq 0$

[4] $\qquad g_0^0 * g_2^4 = -8f_1 + 2f_{13} + f_{02} - 3f_{03} + 10f_0 \geq 0$

[5] $\qquad \overline{g_1}^5 = -f_0 + f_1 - f_2 + f_3 - 4 \geq 0$

By using the fact that for 2-simplicial polytopes the inequality $f_{02} = 3f_2$ holds we can prove the infeasibility of these five inequalities. The following non-negative combination of the inequalities shows the infeasibility. This was obtained by Fourier-Motzkin Elimination.

$$3 * [1] + 3 * [2] + [3] + 2 * [4] + 24 * [5] = -4f_0 - 10f_1 - 6f_3 - 96 < 0 \qquad \square$$

Theorem 4.3. *Every 2-simplicial d-polytope $(d \geq 7)$ has a 3-face with less than 7 vertices.*

Proof. Again it suffices to prove the theorem for 7-polytopes. Assume that every 3-face of a 2-simplicial 7-polytope has 7 or more vertices (inequality $f_0^3 - 7 \geq 0$ in the bottom interval $[-1, 3]$) and that every 2-face is triangular (inequality $3 - f_0^2 \geq 0$ in the interval $[-1, 2]$). Note that $g_1^2 = f_0^2 - 3 \geq 0$ and therefore $f_0^2 = 3$. Consider the following 15 inequalities for 7-polytopes obtained by convolutions of the g-numbers, their duals and the added inequalities. The theorem follows again from the infeasibility of this system of linear inequalities.

[1] $\quad (3 - f_0) * g_0^0 * g_1^2 * g_0^0 = 18f_{03} - 36f_3 + 18f_{24} + 3f_{035} - 6f_{35} - 6f_{13}$
$$- 6f_{024} - f_{135} \geq 0$$

[2] $\quad (f_0 - 7) * g_1^2 * g_0^0 = -6f_{03} - f_{035} + 3f_{024} + 42f_3 + 7f_{35} - 21f_{24}$
$$+ 15f_{14} - 15f_{04} \geq 0$$

[3] $g_0^1 * g_1^4 * g_0^0 = -f_{024} + f_{025} - 10f_1 - 3f_{13} + 5f_{14} - 5f_{15}$
            $+ 5f_{02} \geq 0$

[4] $g_0^1 * g_2^4 * g_0^0 = 3f_{024} - f_{025} + 20f_1 + 4f_{13} - 10f_{14} + 4f_{15}$
            $- 10f_{02} \geq 0$

[5] $g_1^6 * g_0^0 = -14 + 7f_0 + 7f_2 - 7f_3 + 7f_4 - 7f_5 - f_{02}$
            $+ f_{03} - f_{04} + f_{05} - 5f_1 \geq 0$

[6] $(3 - f_0) * g_1^2 * g_0^1 = 6f_{35} - 3f_{035} + f_{135} - 9f_{25} + 3f_{025} \geq 0$

[7] $(f_0 - 7) * g_0^0 * g_1^2 = -3f_{024} + 2f_{035} + 15f_{04} - 15f_{14} + 21f_{24}$
            $- 14f_{35} \geq 0$

[8] $(3 - f_0) * g_0^1 * g_1^2 = f_{135} - 3f_{035} + 6f_{35} - 9f_{24} + 3f_{024} \geq 0$

[9] $g_0^1 * g_1^2 * g_1^2 = -3f_{024} - 6f_{15} - f_{135} + 3f_{025} + 9f_{14} \geq 0$

[10] $g_0^4 * g_1^2 = 2f_5 - f_{05} + f_{15} - f_{25} + f_{35} - 3f_4 \geq 0$

[11] $(f_0 - 7) * g_0^3 = f_{03} - 7f_3 \geq 0$

[12] $g_0^2 * g_2^4 = -8f_3 + 4f_{03} - 4f_{13} + 2f_{35} - f_{035} + f_{135}$
            $+ f_{24} - 3f_{25} + 10f_2 \geq 0$

[13] $g_1^2 * g_2^4 = 24f_3 - 12f_{03} - 6f_{35} + 3f_{035} + 4f_{13} - f_{135}$
            $+ f_{024} - 3f_{025} + 10f_{02} - 3f_{24} + 9f_{25}$
            $- 30f_2 \geq 0$

[14] $(3 - f_0) * g_0^4 = -f_{02} + 3f_2 \geq 0$

[15] $g_0^0 * \overline{g_1}^{-6} = -f_{02} + f_{03} - f_{04} + f_{05} + 2f_1 - 7f_0 \geq 0$

The following positive combination of the 15 inequalities obtained by FLAGTOOL proves the infeasibility of the system.

$112*[1] + 84*[2] + 84*[3] + 105*[4] + 540*[5] + 112*[6] + 168*[7] + 336*[8] + 210*[9] + 1260*$
$[10] + 252*[11] + 189*[12] + 315*[13] + 1260*[14] + 720*[15] = -1260f_5 - 1260f_0 - 7560 < 0$
$\square$

Theorems 4.2 and 4.3 were obtained by direct support of FLAGTOOL. The added inequalities, as part of the user input for our program, are $f_0 - 3 \leq 0$ in the bottom interval $[-1, 2]$ and $f_0 - x_d \geq 0$ in the bottom interval $[-1, 3]$. The quantity x_d is a conjectured lower bound for the number of vertices of a 3-face. Several tests with different values for the dimension d and the quantity x_d lead to the following lower bounds for x_d, which are sufficient to show the infeasibility of the system of n inequalities. Note that in all cases the numbers g_0, g_1, g_2 are sufficient to prove the theorems and so they are not restricted to rational polytopes. (The first and third rows of the table are just Theorems 5 and 6 above.)

d	x_d	use of g_i	n
5	8	g_2	16
6	8	g_2	32
7	7	g_2	65
8	7	g_2	122
9	7	g_2	227
10	7	g_2	424

Similar work with FLAGTOOL shows that every 3-simplicial d-polytope ($d \geq$ 7) has a small 4-face.

The added inequalities are $f_0 - 4 \leq 0$ in the bottom interval $[-1,3]$ and $f_0 - x_d \geq 0$ in the bottom interval $[-1,4]$. The following lower bounds for the vertex number x_d of a 4-face for which the infeasibility holds, the use of the g-numbers and the number n of generated inequalities were obtained.

d	x_d	use of g_i	n
7	10	g_2	60
8	10	g_2	111
9	9	g_2	206
10	9	g_2	382

4.4. Low dimensional quotients with few vertices

It is a classical result that every 3-polytope or its dual has a triangular 2-face. It follows that every d-polytope ($d \geq 3$) has a 2-quotient Q which is a triangle. Indeed if a 3-polytope P and his dual both has no triangular face then we will obtain that

[1] $$(f_0 - 4) * g_0^0 = -8 + 4f_0 - 2f_1 \geq 0$$
[2] $$g_0^0 * (f_0 - 4) = 2f_1 - 4f_0 \geq 0$$

$$[1] + [2] = -8 < 0$$

In what follows we will prove that higher dimensional analogous of this fact hold as well. We show that high dimensional polytopes always have small 3-dimensional quotients in some interval of the face lattice. The most important result is that every d-polytope ($d \geq 9$) has a 3-quotient which is a simplex. For lower dimensions we show that every d-polytope ($5 \leq d \leq 8$) has a 3-quotient with less than x_d vertices. These results were obtained by adding the 3-dimensional inequality $f_0 - x_d \geq 0$ to all intervals $[-1,3] \ldots [d-4,d]$. The lower bounds for the quantity x_d, the use of the g-numbers and the number n of nonnegative convolutions (inequalities) produced by FLAGTOOL are as follows:

d	x_d	use of g_i	n
5	8	g_1	14
6	7	g_1	28
7	6	g_1	55
8	6	g_1	103
9	5	g_2	243

In particular,

Theorem 4.4. *Every d-polytope (d ≥ 9) has a 3-quotient which is a simplex.*

In a similar way we prove that every d-polytope $(7 \leq d \leq 9)$ has a 4-quotient with less than x_d vertices. The values for x_d, the use of the g-numbers and the number n of nonnegative d-forms produced by FLAGTOOL are as follows:

d	x_d	use of g_i	n
7	16	g_2	61
8	13	g_2	112
9	10	g_2	210

4.5. Low dimensional "small" quotients in prescribed locations

Note that Conjecture 4.2 would imply that if d is large enough we can always find a k-dimensional quotient of the form G/F which is a simplex, where the dimension e of F is specified and $e \neq -1, d - k - 1$. In this section we show that 2- and 3-dimensional quotients with small number of vertices appear in certain specified intervals of a high dimensional polytope.

A consequence of the fact that every 3-polytope or its dual has a triangular 2-face and of the fact that the union of two adjacent 2-dimensional intervals are representing a 3-dimensional interval is the following corollary.

Corollary 4.5. *Every d-polytope (d ≥ 3) has a triangle as a 2-quotient either in the interval $[e, e + 3]$ or in the interval $[e + 1, e + 4]$ $(-1 \leq e \leq d - 4)$.*

In what follows we present similar, but new results concerning triangular 2-quotients in certain locations. Again all results were obtained by direct support of FLAGTOOL.

Theorem 4.6. *Every d-polytope (d ≥ 6) has a triangular 2-quotient either in the interval $[0, 3]$ or in the interval $[2, 5]$. In particular, every 6-polytope or its dual has a 3-face with a 3-valent vertex.*

Theorem 4.7. *Every 7-polytope has a triangle as a 2-quotient of a 1-face in a 4-face, i. e., it has a 4-face with an edge that is contained in three 3-faces.*

Moreover we prove that there always exist small 3-quotients in certain interesting locations. For example we show that every d-polytope $(7 \leq d \leq 9)$ has a small 3-face or its dual has a small 3-face with less than x_d vertices. The added inequalities for FLAGTOOL are $f_0 - x_d \leq 0$ in the bottom interval $[-1, 3]$ and in the top interval $[d-4, d]$. The quantities x_d, the use of the g-numbers and the number n of nonnegative convolutions (inequalities) produced by FLAGTOOL are as follows:

d	x_d	use of g_i	n
7	17	g_2	62
8	55	g_2	116
9	21	g_2	194

Next we show that from a certain dimension there exist small 3-quotients in the intervals $[0, 4]$ and $[1, 5]$. Several tests with FLAGTOOL by adding the inequality $f_0 - x_d \leq 0$ in the interval $[0, 4]$ lead to the following lower bounds x_d for which the infeasibility of the n inequalities obtained by convolutions of the g-numbers and the added inequality holds.

d	x_d	use of g_i	n
5	12	g_2	14
6	12	g_2	28
7	10	g_2	57
8	10	g_2	104
9	10	g_2	172

For the interval $[1, 5]$ the result is as follows.

d	x_d	use of g_i	n
6	12	g_2	28
7	8	g_2	56
8	8	g_2	103
9	8	g_2	168

4.6. Low dimensional quotients with small g_2

In the previous sections low dimensional quotients with small g_1 were considered. We will now consider quotients with small g_2.

Theorem 4.8. *Every d-polytope $(d \geq 7)$ has a 4-quotient Q such that $g_2(Q) = 0$.*

5. Other possible applications and extensions

5.1. Non-linear inequalities and linear consequences

The nonlinear relations among g-numbers which are known to hold for simplicial polytopes are conjectured for general polytopes. Other nonlinear inequalities were recently proved. Convolutions still apply, what can be derived from these inequalities? Do they imply *linear* inequalities for the flag numbers?

For simplicial d-polytopes (and more generally, for subcomplexes of their boundary complexes) one can derive quite sharp upper bounds for f_i in terms of f_{i-1}. (These bounds called the generalized upper bound inequalities are attained for cyclic polytopes). See [Kal91]. For $i \geq [d/2] + 1$ these bounds are linear. Are these bounds continue to apply for the non-simplicial case? This is conjectured to be true in [Kal91]. This conjecture would imply a positive answer to Barany's question (Section 3.3) as well as the following remarkable extension of Björner's partial unimodality results for simplicial polytopes [Bj94]:

Conjecture 5.1. *The face numbers f_i of d-polytopes are non-decreasing for $i \leq [(d+3)]/4$ and nonincreasing for $i \geq [3(d-1)/4]$.*

The result of Braden and MacPherson which we are going now to discuss may be relevant or even the key for proving such a conjecture.

There are various relations involving flag numbers of a polytope P and those of a specific face F and the quotient P/F. Braden and MacPherson [BrMP] proved for rational polytopes that

$$g_P(x) \geq g_F(x) \times g_{P/F}(x).$$

(Here the inequality means that all coefficients of the polynomial on the left hand side are at least as large as those in the right hand side.) FLAGTOOL does not involve such relations. Can they be added to the picture? For simplicial complexes relations between face numbers of a complex and its links are fundamental in proving nonlinear relations [McM70, BjK91]. The Braden-MacPherson inequalities already have various interesting applications [Kal88, Bay98, BiEh]. Bayer found sharp form of the upper bound theorem for general polytopes and Billera and Ehrenborg used the Braden-MacPherson result and their own monotonicity theorem for the cd-index to derive various nonlinear inequalities. It seems that there is much yet to be explored.

5.2. Special classes of polytopes

Simplicial and cubical polytopes

The face numbers of simplicial polytopes are completely characterized. (Although finding interesting combinatorial consequences from this characterization is still a challenge, see e. g. [Bj94].) Perhaps it is the right time to introduce and study more delicate numerical invariants for them. (For some ideas see [Gr70].) E. g., for simplicial 4-polytopes we can consider the number of pairs of facets which has an edge in common. (Or, similarly, we can study the f-vector of the deleted join of the polytope.)

The knowledge of cubical polytopes is much less complete (see [Ad96, BBC97, JZ]). For both simplicial and cubical polytopes flag numbers are determined by face numbers and the affine space spanned by face numbers is of dimension $[d/2]$.

In the cubical case there are analogs of the g_i's introduced by Adin [Ad96] who conjectured them to be non-negative. It would be interesting to find combinatorial consequences from the nonnegativity of the g_i for the links (and possibly also Adin's conjecture) e. g., finding an analog of Nikulin's theorem [Nik86, Theorem C] or showing that dual-to-cubical d-polytopes always have a e-dimensional face which is a simplex (where e tends to infinity with d).

Quasi simplicial polytopes

Quasi simplicial polytopes are polytopes all whose faces are simplicial. For these polytopes the face numbers determine all flag numbers [Kal88] and the affine space of face numbers is $d - 1$. It seems that finding the linear inequalities for face-numbers for such polytopes which can be derived from the nonnegativity of the g_i's can be done automatically for much higher dimensions ($d \leq 100$ seems realistic).

As pointed out by Anders Björner such a study can contribute to the classification of hyperbolic reflection groups, see [Nik86, Kho86]. The results obtained so far heavily use Nikulin's theorem [Nik86, Thm. C] as well as its extension by Khovanskii to the quasi-simplicial case. For this application we need to consider only the restricted class of polytopes whose facets are Cartesian products of simplices.

k-simplicial $(d - k)$-simple d-polytopes

Recall that a polytope P is called k-simplicial if all its k-faces are simplices. P is k-simple if P^* is k-simplicial. If P is a k-simplicial, r-simple d-polytope ($k, r, \leq d$) and $k + r > d$ then P is a simplex. Finding k-simplicial $(d - k)$-simple d-polytopes (for $k, d - k > 1$), apart from the simplex itself, is of great interest. No such example is known for $k, d - k \geq 4$. The convex hulls of middle points of the edges of the simplex (the 2-hypersimplex) are 2-simplicial $(d - 2)$-simple polytopes. The convex hull of the even vertices of the d-cube are 3-simplicial $(d - 3)$-simple. (These examples and others were worked out by Perles who may have had also a single example for $k = d - k = 4$ that was forgotten.) It would be interesting to understand the linear relations among flag vectors of such polytopes.

Facet-forming polytopes

A d-polytope P is a facet forming polytope if there is a $(d + 1)$-polytope all whose facets are isomorphic to P. By eliminating variables we can, in principle, determine the average behavior of flag numbers of facets of a d-polytope (and more generally the behavior of the averages of flag numbers for e-dimensional faces of d-polytope). Clearly every such inequality will apply to facet-forming polytopes. It turns out that for $d = 4, 5$ no new inequalities are obtained. More sophisticated applications of the inequalities in $\mathcal{Q}_5$ can be used to show that certain 4-polytopes are not facet-forming, see [Kal90]. It would be interesting to extend these results to higher dimension.

Barycentric subdivisions

Another interesting direction is to try to determine all linear relations between face numbers of barycentric subdivisions of d-polytopes. Recall that the number $f_k(BP)$ of the barycentric subdivision BP of P is simply the sum of $f_S(P)$ for all sets S of cardinality $k + 1$. Again, this is a question about the projection of $\mathcal{Q}'_d$ to some small dimensional subspace. This direction was not worked out yet.

Zonotopes

Understanding flag numbers of zonotopes is of great interest. There are several famous open problems and some advances seem to be related to the kind of arguments used here. Billera, Ehrenborg and Readdy [BER97] showed that the affine space spanned by flag number of d-dimensional zonotopes is the entire space spanned by flag numbers of d-polytopes. (In other words, they proved the existence of a basis of polytopes all whose members are zonotopes. They did not construct explicitly such a basis.)

Other posets

There is much activity concerning the combinatorics and the linear relation of flag numbers of various classes of ranked posets [BayH, BilH, BilH']. The generalized Dehn-Sommerville inequalities apply to all Eulerian posets while the inequality $g_1 \geq 0$ aplly to all relatively complemented lattices. Thus, both these properties apply to general Eulerian lattices and FLAGTOOL can thus be useful to their study. (We do not know, for example, if the conjectures of Section 4.1 may apply to arbitrary Eulerian lattices.)

5.3. Extensions of FLAGTOOL

High dimensions

It may be useful to try to extend FLAGTOOL to higher dimensions. Since the number of variables and inequalities grows rapidly and the resulting LP problems are not sparse we cannot expect too much but extending the program up to 20 dimensions may be feasible. As we already mentioned, for restricted classes of polytopes (quasisimplicial, for example) we may hope for a similar program applicable in much larger dimensions.

Further automation

FLAGTOOL is used to automate certain combinatorial arguments which were done "by hand" in a few papers and with some computer support in another [Ba74, Bay87, Kal90].

Let $\mathcal{G}$ be a specific system of linear inequalities for flag numbers of k-polytopes. Most of the theorems proved using FLAGTOOL are of the following form:

Every d-polytope contains a k-dimensional quotient (possibly in a prescribed location) which satisfies the inequalities in $\mathcal{G}$.

The current modus operandi (see Figure 1) was that we tested various conjectures of this kind using FLAGTOOL, but here also further automation seems possible.

Problem 5.2. *Find automatically (all the) theorems of this form that are derived from the known linear inequalities.*

Note that this problem is not identical to finding all flag-number inequalities which hold on average for k-dimensional quotients in prescribed locations.

Using other types of reasoning

There are arguments which are similar to those we use but use additional ingredients. For example: studying the behavior of certain face number relations along a shelling process of the polytope. See for example [Ba80, BlBl90]. Can these arguments be automated (and thus systematically extended) too?

Blind and Blind [BlBl90] proved that every d-polytope with no triangular 2-faces must contain at least as many k-faces as the d-dimensional cube. This is a type of theorem that could have been derived using FLAGTOOL but it does not follow from the known flag number inequalities. The proof of Blind and Blind contains further ingredients.

Challenge 5.3. *Automate (and extend) the theorem of Blind and Blind.*

Challenge 5.4. *Automate (and extend) the proofs of* [BBMM90].

Seymour's arguments [Sey82] in connection with the points-lines-planes conjecture have some similar flavor to the arguments used here.

Challenge 5.5. *Automate (and extend) Seymour's theorems.*

6. Conclusion

FLAGTOOL can serve as a useful tool for proving theorems concerning the combinatorial structure of polytopes of dimension $d \leq 10$ and for testing and making conjectures for arbitrary polytopes. At present the proofs obtained from FLAGTOOL do not seem to give much insight (to humans) about the theorems and, in particular, do not supply a recipe for extensions to higher dimensions.

7. Appendix

FLAGTOOL is a computer program implementing the ideas described in this paper; see also [Mei94]. The code may be obtained from the second author on request.

7.1. Description of the available tools

After starting FLAGTOOL, a menu whose topics correspond to the available tools appears on the screen. After executing a tool the program returns to the menu. Most of the tools are simple I/O-programs, i. e. they transfer data from or to a file or screen. Other tools like (5), (8), (13) and (18) require nontrivial data structures and algorithms. Here is a complete enumeration of the available tools.

(1) DIMENSION

This simple tool is used to change the current working dimension d and provides the new basic input for d. FLAGTOOL accepts values from dimension 3 up to dimension 10 (this can easily be raised to higher dimensions).

(2) GMAX

This changes the use of the g-numbers. The maximum possible value depends on the current working dimension. If this value is less than three, i. e. only g_0, g_1 and g_2 are used in convolutions, results hold for all polytopes, not only for rational polytopes (see section 1.6).

(3) ADD

This tool is used for adding a new inequality for proper faces or quotients of d-polytopes to the system (maybe the negation of a conjectured inequality) and for specifying the dimension of this inequality and the intervals in which this inequality appears.

(4) DELETE

Delete is used for removing added inequalities (with ADD) and specified intervals.

(5) MAKE

This tool computes a system A of linear inequalities by convolution of the g-inequalities and the added inequalities in specified intervals. Every inequality is expressed in terms of special flag numbers. See section 4.4 of [Mei94] for data structures and algorithms.

(6) INADD

This adds a new inequality in the present working dimension to a current system A of linear inequalities, which was created by the MAKE-tool by convolutions of the g-numbers and added inequalities in specified intervals and changed by tool (6), (7), (8) or (9).

(7) INDEL

This serves for deleting an inequality from a current system A of linear inequalities.

(8) ELIM

This tool eliminates a special flag number from the current system A of linear inequalities.

(9) DSELIM

This eliminates a special flag number from the current system A of linear inequalities using only Generalized Dehn Sommerville Equations.

(10) MPS

This tool creates an mps-file (LP input format) which corresponds either to the current system of linear inequalities (an objective function in terms of special flag numbers can be specified) or to the dual problem.

(11) SAVE

A current system of linear inequalities is saved on a file.

(12) FETCH

This reads a system of linear inequalities from a file created by the SAVE-tool.

(13) DISPLAY

This tool is used for displaying the current input or more information about flag numbers. The following options are available.

  (a) STATUS
      Display the current input of FLAGTOOL (working dimension, use of
      the g-numbers and added inequalities).
  (b) SYSTEM
      Display the current system A of inequalities.
  (c) G-NUM
      Display the g-numbers and their duals up to the present working di-
      mension.
  (c) DEHN
      Display the expression of the flag numbers by the special flag numbers
      as a solution of the Generalized Dehn Sommerville Equations.
(14) READ
     This reads input for FLAGTOOL from a file.
(15) WRITE
     This writes the current FLAGTOOL input to a file.
(16) SOLVE
     This tool is available only if FLAGTOOL is linked with the linear program-
     ming solver $CPLEX^{TM}$ [1]. It computes the objective function value, if it is
     specified, or simply tests the infeasibility of an mps-file created by FLAG-
     TOOL.
(17) RED
     This tool removes redundant inequalities from a given system A of linear
     inequalities produced, for example, by the ELIM-tool. Only if FLAGTOOL
     is linked with the LP-solver $CPLEX^{TM}$ RED-tool is available.
(18) CD
     This tool computes the coefficients in the *cd index* (see section 1.10) for the
     present working dimension in terms of special flag numbers.
(19) DUAL
     For a current system of linear inequalities the system of the corresponding
     dual inequalities is computed.
(19) HELP
     This provides the user with online information about the program, the tools
     and the interpretation of the output.

7.2. Working with FLAGTOOL

A short demonstration into how FLAGTOOL works is given by explaining a typical
FLAGTOOL-session that proves Kalai's result about the existence of small 2-faces
in d-polytopes $(d \geq 5)$.

FLAGTOOL starts with the following menu. Note, that the following con-
ventions are used to distinguish computer output from user input. All output
produced by the computer will appear in `typewriter-like` font. Text entered by
the user will appear in *italic* font.

---

[1]$CPLEX^{TM}$ is a registered trademark of CPLEX OPTIMIZATION INC.

```
Welcome to FLAGTOOL!

Here is a list of available commands.
Type 'help' followed by a command name for more
information on commands, for example 'helpadd'.

(dim)ension     set or change the working dimension
(gm)ax          set or change the use of the $g$-numbers
(ad)d           add an inequality for proper faces or quotients
(de)lete        delete one or more added inequalities
(ma)ke          compute a system of inequalities by convolution
(ina)dd         add an inequality to a current system
(ind)el         delete inequalities from a current system
(el)im          eliminate a flag from a current system
(ds)elim        eliminate a flag by using Dehn Sommerville
(mp)s           create an mps-file
(sa)ve          write a current system to a file
(fe)tch         get a current system from a file
(dis)play       display the input, the current system or
                more information about flag numbers
(re)ad          read input from a file
(wr)ite         write input to a file
(so)lve         solve an mps-file with CPLEX
(red)undant     remove redundancy from a current system
(cd)index       compute the coefficients in the cd index
(du)al          compute the dual system of inequalities
(com)mands      list the FLAGTOOL commands
(qu)it          leave FLAGTOOL

Commands may be executed by entering the command name (or
at least the letters in the bracket) and FLAGTOOL will
prompt you for additional required information.
```

First we have to set the working dimension and the use of the g-numbers. Note that g_2 is needed for the proof, i. e. FLAGTOOL cannot prove Kalai's theorem only with g_1.

```
FLAGTOOL> dim

Present value for the working dimension: 3
New value for the working dimension: 5

Okay, new value for the working dimension: 5
```

FLAGTOOL> *gm*

```
Present value for the use of the g-numbers: 1
New value for the use of the g-numbers: 2
```

```
New value for the use of the g-numbers: 2
```

In order to prove that every 5-polytope has a 2-face with less than 5 vertices we have to add the inequality $f_0^2 - 5 \geq 0$ to the bottom interval $[-1, 2]$ ('add' command), convolve this inequality with the numbers g_i^k and $\overline{g}_i^{-k}$ ($0 \leq k \leq 5, 0 \leq i \leq \lfloor k/2 \rfloor$) ('make' command), create an mps-file ('mps' command) and solve the corresponding linear program ('solve' command).

FLAGTOOL> *add*

```
There are 0 added inequalities.
```

```
Enter the dimension of the new inequality: 2
```

```
Enter the new inequality: f0-5
```

```
Enter the interval(s) in which it appears: [-1,2]
```

```
Inequality added!
```

FLAGTOOL> *make*

```
Note, that you have added inequalities for proper
faces or quotients!
```

```
Flagtool computes a system of inequalities, please wait ...
```

```
Current system A with 15 inequalities created!
```

FLAGTOOL> *mps*

```
Enter '1' (dual) or '0' (primal): 0
```

```
Enter '1' (min) or '0' (max): 1
```

```
Enter the objective function: 0
```

```
Enter a name for the new mps-file : kalai5.mps
```

```
File kalai5.mps created!
```

FLAGTOOL> *solve*

```
Enter the name for the mps-file : kalai5.mps
```

```
problem is infeasible!
```

Before the session ends we have a look at the current system of linear inequalities by using the '(dis)play' command.

FLAGTOOL> *dis*

```
Display options :

(sta)tus          display the current input
(sys)tem          display the currrent system of inequalities
                  and their meanings
(g_n)um           display the g-numbers up to the present
                  working dimension
(deh)n            display the Dehn Sommerville Equations
                  for the specified working-dimension

Display what: sys
Working dimension: 5

      1   -6f1-f13+3f02 = (G_0_1)*(G_1_2)*(G_0_0)
      2   -10+5f0+5f2-5f3-f02+f03-3f1 = (G_1_4)*(G_0_0)
      3   -10+5f0-5f1+5f2-3f3 = (G^_1_4)*(G_0_0)
      4   20-10f0-10f2-f03+4f3+4f1+3f02 = (G_2_4)*(G_0_0)
      5   2f13-3f03 = (G_0_0)*(G_1_2)*(G_0_1)
      6   2f3-f03+f13-3f2 = (G_0_2)*(G_1_2)
      7   f02-3f2 = (G_1_2)*(G_0_2)
      8   -6f3+3f03-f13-3f02+9f2 = (G_1_2)*(G_1_2)
      9   f02-5f2 = (-5+f0)*(G_0_2)
     10   -10f3+5f03-3f13-3f02+15f2 = (-5+f0)*(G_1_2)
     11   2f1-5f0 = (G_0_0)*(G_1_4)
     12   -f02+f03+2f1-5f0 = (G_0_0)*(G^_1_4)
     13   -8f1+2f13+f02-3f03+10f0 = (G_0_0)*(G_2_4)
     14   -6+f0 = (G_1_5)
     15   -f0+f1-f2+f3-4 = (G^_1_5)

FLAGTOOL> quit
```

The infeasibility of the system of 15 inequalities and thus the correctness of the theorem can be proved by hand. Consider the six nonnegative 5-forms 1, 5, 9, 10, 13 and 15. The following nonnegative combination of these six inequalities results in an inequality which is strictly less than zero and therefore the infeasibility is proved.

$$(-6f_1 - f_{13} + 3f_{02})$$
$$+ 4 * (2f_{13} - 3f_{03})$$
$$+ 8 * (f_{02} - 5f_2)$$
$$+ 3 * (-10f_3 + 5f_{03} - 3f_{13} - 3f_{02} + 15f_2)$$
$$+ (-8f_1 + 2f_{13} - 3f_{03} + f_{02} + 10f_0)$$
$$+ 10 * (-f_0 + f_1 - f_2 + f_3 - 4)$$
$$= -20f_0 - 5f_1 - 4f_2 - 40 < 0$$

7.3. Known flag number inequalities for d-polytopes, $d \leq 6$

We describe now (nonredundant) d-forms (linear combinations of flag numbers of d-polytopes) which are known to be nonnegative, for $d = 3, 4, 5, 6$.

Dimension 3

1      $g_1^2 * g_0^0 = -6 + 3f_0 - f_1$

2      $g_0^0 * g_1^2 = 2f_1 - 3f_0$

Dimension 4

1      $g_0^0 * g_1^2 * g_0^0 = -6f_0 - f_{02} + 6f_1$

2      $g_1^2 * g_0^1 = f_{02} - 3f_2$

3      $g_0^1 * g_1^2 = f_{02} - 3f_1$

4      $g_1^4 = -5 + f_0$

5      $\overline{g_1}^4 = f_0 - f_1 + f_2 - 5$

6      $g_2^4 = 10 - 4f_0 + f_{02} - 3f_2 + f_1$

Dimension 5

1      $g_0^1 * g_1^2 * g_0^0 = -6f_1 - f_{13} + 3f_{02}$

2      $g_1^4 * g_0^0 = -10 + 5f_0 + 5f_2 - 5f_3 - f_{02} + f_{03} - 3f_1$

3      $\overline{g_1}^4 * g_0^0 = -10 + 5f_0 - 5f_1 + 5f_2 - 3f_3$

4      $g_2^4 * g_0^0 = 20 - 10f_0 - 10f_2 - f_{03} + 4f_3 + 4f_1 + 3f_{02}$

5      $g_0^0 * g_1^2 * g_0^1 = 2f_{13} - 3f_{03}$

6      $g_0^2 * g_1^2 = 2f_3 - f_{03} + f_{13} - 3f_2$

7      $g_1^2 * g_0^2 = f_{02} - 3f_2$

8      $g_1^2 * g_1^2 = -6f_3 + 3f_{03} - f_{13} - 3f_{02} + 9f_2$

9      $g_0^0 * g_1^4 = 2f_1 - 5f_0$

10     $g_0^0 * \overline{g_1}^4 = -f_{02} + f_{03} + 2f_1 - 5f_0$

11     $g_0^0 * g_2^4 = -8f_1 + 2f_{13} + f_{02} - 3f_{03} + 10f_0$

12     $g_1^5 = -6 + f_0$

13     $\overline{g_1}^5 = -f_0 + f_1 - f_2 + f_3 - 4$

Dimension 6

1      $g_0^2 * g_1^2 * g_0^0 = -6f_2 - f_{24} + 3f_{13} - 3f_{03} + 6f_3$

2      $g_1^2 * g_1^2 * g_0^0 = -6f_{02} - f_{024} + 18f_2 + 3f_{24} - 3f_{13} + 9f_{03} - 18f_3$

3      $g_0^0 * g_1^4 * g_0^0 = -2f_{13} + 2f_{14} - 10f_0 - 3f_{02} + 5f_{03} - 5f_{04} + 10f_1$

$$4 \quad g_0^0 * \overline{g_1}^4 * g_0^0 = -10f_0 - 5f_{02} + 5f_{03} - 3f_{04} + 10f_1$$

$$5 \quad g_0^0 * g_2^4 * g_0^0 = 6f_{13} - 2f_{14} + 20f_0 + 4f_{02} - 10f_{03} + 4f_{04} - 20f_1$$

$$6 \quad g_1^5 * g_0^0 = -6f_2 + 6f_3 - 6f_4 - 4f_0 + f_{02} - f_{03} + f_{04} + 4f_1$$

$$7 \quad \overline{g_1}^5 * g_0^0 = -6f_0 + 6f_1 - 6f_2 + 6f_3 - 4f_4$$

$$8 \quad g_0^1 * g_1^2 * g_0^1 = f_{024} - 3f_{14}$$

$$9 \quad g_1^4 * g_0^1 = f_{04} - 5f_4$$

$$10 \quad \overline{g_1}^4 * g_0^1 = f_{04} - f_{14} + f_{24} - 5f_4$$

$$11 \quad g_2^4 * g_0^1 = f_{14} - 3f_{24} + f_{024} - 4f_{04} + 10f_4$$

$$12 \quad g_1^2 * g_0^0 * g_1^2 = 2f_{024} + 18f_3 - 9f_{03} + 3f_{13} - 6f_{24}$$

$$13 \quad g_0^0 * g_1^2 * g_0^2 = 2f_{13} - 3f_{03}$$

$$14 \quad g_0^0 * g_1^2 * g_1^2 = -6f_{13} - 6f_{04} - f_{024} + 6f_{14} + 9f_{03}$$

$$15 \quad g_0^3 * g_1^2 = f_{04} - f_{14} + f_{24} - 3f_3$$

$$16 \quad g_1^2 * g_0^3 = f_{02} - 3f_2$$

$$17 \quad g_0^1 * g_1^4 = f_{02} - 5f_1$$

$$18 \quad g_0^1 * \overline{g_1}^4 = -f_{13} + f_{14} + f_{02} - 5f_1$$

$$19 \quad g_0^1 * g_2^4 = -4f_{02} + f_{024} + f_{13} - 3f_{14} + 10f_1$$

$$20 \quad g_0^0 * g_1^5 = 2f_1 - 6f_0$$

$$21 \quad g_0^0 * \overline{g_1}^5 = f_{02} - f_{03} + f_{04} - 2f_1 - 4f_0$$

$$22 \quad g_1^6 = -7 + f_0$$

$$23 \quad \overline{g_1}^6 = f_0 - f_1 + f_2 - f_3 + f_4 - 7$$

$$24 \quad g_2^6 = 21 - 6f_0 + f_{02} - 3f_2 + f_1$$

$$25 \quad \overline{g_2}^6 = f_{24} - f_{14} + f_{04} - 6f_0 + 6f_1 - 6f_2 + 21 + 3f_3 - 5f_4$$

$$26 \quad g_3^6 = -35 + 15f_0 - 4f_{02} - 5f_1 + f_{14} - 3f_{24} + f_{024} - 4f_{04} + 10f_4$$

$$+ f_{03} - 4f_3 + 13f_2 \text{ (nonnegativity is known}$$

$$\text{only for rational polytopes)}$$

$$27 \quad Added = f_2 - 35 \text{ (see Section 1.9)}$$

$$28 \quad Added = f_3 - 35$$

References

[Ad96] R. Adin: A new cubical h-vector, *Discrete Math.* **157** (1996), 3–14.

[BBC97] E. Babson, L. J. Billera and C. Chan: Neighborly cubical spheres and a cubical lower bound conjecture, *Israel J. Math.* **102** (1997), 297–315.

[Ba71] D. W. Barnette: A proof of the lower bound conjecture for convex polytopes, *Pacific J. Math.* **46** (1971), 349–354.

[Ba74] D. W. Barnette: The projection of the f-vectors of 4-polytopes onto the (E, S)-plane, *Discrete Math.* **10** (1974), 201–216.

[Ba80] D. W. Barnette: Nonfacets for shellable spheres, *Israel J. Math.* **35** (1980), 286–288.

[BaBo96] V. V. Batyrev and L. A. Borisov: Mirror duality and string theoretic Hodge numbers, *Invent. Math.* **126** (1996), 183–203.

[Bay87] M. M. Bayer: The extended f-vectors of 4-polytopes, *J. Combinat. Theory Ser. A* **44** (1987), 141–151.

[Bay93] M. M. Bayer: Equidecomposable and weakly neighborly polytopes, *Israel J. Math.* **81** (1993), 301–320.

[Bay94] M. M. Bayer: Face numbers and subdivisions of convex polytopes, in: "Polytopes: Abstract, Convex and Computational" (T. Bisztriczky et al., eds.) Kluwer 1994, 155–172.

[Bay98] M. M. Bayer: An upper bound theorem for rational polytopes, *J. Combinat. Theory Ser. A* **83** (1998), 141–145.

[BayB85] M. M. Bayer and L. J. Billera: Generalized Dehn-Sommerville relation for polytopes, spheres and Eulerian partially ordered sets, *Invent. Math.* **79** (1985), 143–157.

[BayE] M. M. Bayer and R. Ehrenborg: On the toric h-vectors of partially ordered sets, *Transactions Amer. Math. Soc.*, to appear.

[BayH] M. M. Bayer and G. Hetyei: Flag vectors of Eulerian partially ordered sets, preprint, 1999.

[BayK91] M. M. Bayer and A. Klapper: A new index for polytopes, *Discrete Comput Geometry* **6** (1991), 33–47.

[BayL93] M. M. Bayer and C. Lee: Convex polytopes, in: *Handbook of Convex Geometry* (P. Gruber and J. Wills, eds.), Vol. A, North-Holland (Elsevier Science Publishers), Amsterdam, 1993, 485–534.

[BBMM90] A. Bezdek, K. Bezdek, E. Makai and P. McMullen: Facets with fewest vertices, *Monatshefte Math.* **109** (1990), 89–96.

[BiBj97] L. J. Billera and A. Björner: Face Numbers of Polytopes and Complexes, in: *Handbook of Discrete and Computational Geometry* (J.E. Goodman, J. O'Rourke, eds.), CRC Press, Boca Raton, New York 1997, 291–310.

[BiLe81] L. J. Billera and C. W. Lee: A proof of the sufficiency of McMullen's conditions for f-vectors of simplicial convex polytopes, *J. Combinat. Theory Ser. A* **31** (1981), 237–255.

[BiEh] L. J. Billera and R. Ehrenborg: Monotonicity of the cd-index for polytopes, *Math. Z.*, **233** (2000), 421–444.

[BER97] L. J. Billera, R. Ehrenborg and M. Readdy: The c-2d-index of oriented matroids *J. Combinat. Theory Ser. A* **80** (1997), 79–105.

[BER98] L. J. Billera, R. Ehrenborg and M. Readdy: The **cd**-index of zonotopes and arrangements *Mathematical Essays in Honor of Gian-Carlo Rota*, B. Sagan and R. Stanley, eds.), Birkhäuser, Boston 1998, 23–40.

[BilH] L. J. Billera and G. Hetyei: Linear inequalities for flags in graded partially ordered sets, *J. Combinat. Theory, Ser. A*, **89** (2000), 77–104.

[BilH'] L. J. Billera and G. Hetyei: Decompositions of partially ordered sets, preprint 1998.

[BilL] L. J. Billera and N. Liu: Noncommutative enumeration in ranked posets, *J. Algebraic Combinatorics*, to appear.

[Bj94] A. Björner: Partial unimodality for f-vectors of simplicial polytopes and spheres, *Contemporary Math.* **178** (1994), 45–54.

[BjK91] A. Björner and G. Kalai: Extended Euler Poincare relations for cell complexes, in: "Applied Geometry and Discrete Mathematics, The Klee Festschrift," *DIMACS Series in Discrete Mathematics and Computer Science* **4** (1991), 81–89.

[BlBl90] G. Blind and R. Blind: Convex polytopes without triangular faces, *Israel J. Math.* **71** (1990), 129–134.

[BrMP] T. Braden and R. D. MacPherson: Intersection homology of toric varieties and a conjecture of Kalai, *Commentarii Math. Helv.*, **74** (1999), 442–455.

[Des] R. Descartes: De Solidorum Elementis, in: *Oeuvres de Descartes*, Vol. 10, 265–276, published by C. Adam and P. Tannery, Paris 1897–1913.

[Gr67] B. Grünbaum: Convex Polytopes, Interscience, London 1967.

[Gr70] B. Grünbaum: Polytopes, graphs and complexes, *Bulletin Amer. Math. Soc.* **97** (1970), 1131–1201.

[Gr72] B. Grünbaum: Arrangements and Spreads, *CBMS Regional Conference Series in Math.* **10**, Amer. Math. Soc., Providence RI 1972.

[JZ] M. Joswig and G. M. Ziegler: Neighborly cubical polytopes, in: "Grünbaum Festschrift" (G. Kalai, V. Klee, eds.), *Discrete Comput. Geometry*, to appear.

[Kal87] G. Kalai: Rigidity and the lower bound theorem I, *Invent. Math. 88* (1987), 125–151.

[Kal88] G. Kalai: A new basis of polytopes, *J. Combinat. Theory, Ser. A* **49** (1988), 191–209.

[Kal90] G. Kalai: On low-dimensional faces that high-dimensional polytopes must have, *Combinatorica 10* (1990), 271–280.

[Kal91] G. Kalai: The diameter of graphs of convex polytopes and f-vector theory, in: "Applied Geometry and Discrete Mathematics, The Klee Festschrift," *DIMACS Series in Discrete Mathematics and Computer Science* **4** (1991), 387–411.

[Kho86] A. G. Khovanskii: Hyperplane sections of polyhedra, toroidal manifolds, and discrete groups in Lobachevskij space, *Funct. Anal. Appl.* **20** (1986), 41–50.

[McM70] P. McMullen: The maximum numbers of faces of a convex polytopes, *Matematika 17* (1970), 179–184.

[McM71] P. McMullen: The numbers of faces of simplicial polytopes, *Israel J. Math.* **9** (1971), 559–570.

[McM93] P. McMullen: On simple polytopes, *Invent. Math.* **113** (1993), 419–444.

[Mei94] G. Meisinger: Flag Numbers and Quotients of Convex Polytopes, Schuch-Verlag, Weiden 1994.

[MKK] G. Meisinger, P. Kleinschmidt and G. Kalai: Three theorems, with computer-aided proofs, on three-dimensional faces and quotients of polytopes, *Discrete Comput. Geometry*, to appear.

[Nik86] V. V. Nikulin: Discrete reflection groups in Lobachevsky spaces and algebraic surfaces, in: *Proc. Int. Congress of Math. Berkeley 1986*, Vol. 1, 654–671.

[Sey82] P. Seymour: On the points-lines-planes conjecture, *J. Combinat. Theory, Ser. B* **33** (1982), 17–26.

[Sta80] R. P. Stanley, The number of faces of simplicial convex polytopes, *Advances Math.* **35** (1980), 236–238.

[Sta87] R. P. Stanley: Generalized *h*-vectors, intersection cohomology of toric varieties, and related results, in: "Commutative Algebra and Combinatorics" (M. Nagata and H. Matsumura, eds.), *Advanced Studies in Pure Mathematics* 11, Kinokuniya, Tokyo, and North-Holland, Amsterdam/New York 1987, 187–213.

[Sta92] R. P. Stanley: Subdivisions and local *h*-vectors, *J. Amer. Math. Soc.* **5** (1992), 805–851.

[Sta94] R. P. Stanley: Combinatorics and Commutative Algebra, Second Edition, Birkhäuser, Boston 1994.

[Sta94'] R. P. Stanley: Flag vectors and the CD-index, *Math. Z.* **216** (1994), 483–499.

[Sta94''] R. P. Stanley: A survey of Eulerian posets, in: "Polytopes: Abstract, Convex and Computational" (T. Bisztriczky et al., eds.) Kluwer 1994, 301–333.

[Zie95] G. M. Ziegler: Lectures on Polytopes, *Graduate Texts in Math.* **152**, Springer-Verlag, New York 1995.

Gil Kalai
Institute of Mathematics
The Hebrew University
91904 Jerusalem
ISRAEL
kalai@math.huji.ac.il

Peter Kleinschmidt
Wirtschaftswissenschaftliche Fakultät
Universität Passau
94032 Passau
GERMANY
kleinsch@winf.uni-passau.de

Günter Meisinger
Serkem GmbH
Watzmannsdorfer Ring 27
94136 Thyrnau
GERMANY
guenter.meisinger@serkem.de

A Census of Flag-vectors of 4-Polytopes

Andrea Höppner and Günter M. Ziegler*

Abstract. How close to a complete description are the known conditions on the flag-vectors of 4-dimensional convex polytopes? We present a computational study (in the POLYMAKE framework) of this question. For small numbers of vertices the conditions are pretty good: only one "impossible flag-vector" satisfies the conditions for $f_0 \leq 7$, and only 12 for $f_0 = 8$.

(1) The set of all f-vectors of (convex, bounded) d-polytopes has dimension $d - 1$: the only linear relation is given by the Euler-Poincaré equation. The corresponding set $\mathcal{F}(\mathcal{P}^d)$ of all the flag-vectors of d-polytopes has dimension $F_d - 1$, where F_d is the d-th Fibonacci number ($F_0 = F_1 = 1$, $F_k = F_{k-1} + F_{k-2}$): their linear relations are given by the generalized Dehn-Sommerville equations of Bayer & Billera [6].

(2) For $d = 3$, both the f-vector and the flag-vector of a 3-polytope is determined by (f_0, f_2). In these coordinates, the set of all f-vectors, and of all flag-vectors of 3-polytopes is completely described by

$$\{(f_0(P), f_2(P)) : P \in \mathcal{P}_3\} \quad = \quad \{(f_0, f_2) \in \mathbb{Z}^2 : \quad \begin{aligned} f_0 &\leq 2f_2 - 4, \\ f_2 &\leq 2f_0 - 4\}, \end{aligned}$$

a classical result of Steinitz [12]. In particular, $\mathcal{F}(\mathcal{P}^3)$ has a finite linear description, so the convex hull $\mathrm{conv}(\mathcal{F}(\mathcal{P}^3))$ is closed, and all the integral points in this set are f-vectors, that is, $\mathrm{conv}(\mathcal{F}(\mathcal{P}^3)) \cap \mathbb{Z}^3 = \mathcal{F}(\mathcal{P}^3)$. A complete characterization (involving non-linear relations) is also available for *simplicial d-polytopes* with the "g-Theorem" of Stanley and Billera & Lee, but not for *general d-polytopes* and any $d \geq 4$.

(3) For $d = 4$, the flag-vector $\mathcal{F}(P)$ of a 4-polytope $P \in \mathcal{P}^4$ is given by

$$\mathcal{F}(P) := (f_0, f_1, f_3, f_{03})$$

together with the linear relations quoted in **(1)**. In these coordinates, the known conditions and restrictions on possible flag-vectors may be summarized as follows (see [5] and [7]):

(3.0) $(f_0, f_1, f_3, f_{03}) \in \{1, 2, 3, \dots\}^4 \subseteq \mathbb{R}^4$

* Supported by a DFG Gerhard-Hess-Forschungsförderungspreis (Zi 475/1-2) and by the German Israeli Foundation (G.I.F.) grant I-0309-146.06/93.

(3.1) Linear inequalities

$$
\begin{array}{lrcl}
(3.1.1) & -f_0 + f_1 + 3f_3 - f_{03} & \leq & 0 \\
(3.1.2) & 2f_0 + f_1 \quad\quad - f_{03} & \leq & 0 \\
(3.1.3) & 3f_0 \quad\quad + 3f_3 - f_{03} & \leq & 10 \\
(3.1.4) & 4f_0 - 4f_1 \quad\quad + f_{03} & \leq & 0 \\
(3.1.5) & f_0 & \geq & 5 \\
(3.1.6) & f_3 & \geq & 5
\end{array}
$$

(3.2) Nonlinear inequalities

$$
\begin{array}{lrcl}
(3.2.1) & 2f_0 - f_1 - 6f_3 + 2f_{03} & \leq & \binom{f_0}{2} \\
(3.2.2) & -5f_0 - f_1 + f_3 + 2f_{03} & \leq & \binom{f_3}{2} \\
(3.2.3) & f_1 - 4f_3 + f_{03} & \leq & \binom{f_0}{2} \\
(3.2.4) & -5f_0 + f_1 + f_3 + f_{03} & \leq & \binom{f_3}{2}
\end{array}
$$

(3.3) Excluded values

$$
\begin{aligned}
\mathcal{F}(P) \notin \{ \ & (7,15,7,32), \quad (7,16,8,35), \\
& (8,17,7,35), \quad (8,18,8,38), \quad (8,18,8,39), \\
& (8,19,10,44), \ (8,20,11,47), \ (8,20,11,48), \\
& (10,21,8,44), \ (11,23,8,47), \ (11,23,8,48) \ \}.
\end{aligned}
$$

(4) Here are some comments on the conditions in **(3)**.

(4.1) The solution set of (3.1) is a 4-dimensional cone whose apex is $\mathcal{F}(\Delta_4) = (5,10,5,20)$. It has six facets and seven extreme rays, see Bayer [5].
The conditions $f_0 \geq 5$ and $f_3 \geq 5$ can be derived from the other conditions:
(3.1.1), (3.1.4), (3.2.1) and (3.2.3) together yield $f_0(f_0 - 3) \geq 10f_3$, so (3.0) yields $f_0 \geq 5$. Dually, one obtains $f_3 \geq 5$.

(4.2) The four quadratic conditions of (3.2) are *concave*, that is, the complements of the solution sets are (strictly) convex. For example, using coordinates that include $x_1 = f_0$ and $x_2 = -2f_1 - 12f_3 + 4f_{03}$, the condition (3.2.1) is $2x_1 + \frac{1}{2}x_2 \leq \binom{x_1}{2}$, that is, $x_2 \leq x_1^2 - 5x_1$. This implies for flag-vectors with equality that no non-trivial convex combination may be a flag-vector. For example, $\mathcal{F}^1 = \mathcal{F}(\Delta_4) = (5,10,5,20)$ and $\mathcal{F}^2 = (8,22,11,50)$ are flag-vectors of simple 4-polytopes with equality in both (3.2.1) and (3.2.3), and hence the convex combinations $\frac{2}{3}\mathcal{F}^2 + \frac{1}{3}\mathcal{F}^2 = (6,14,7,30)$ and $\frac{1}{3}\mathcal{F}^1 + \frac{2}{3}\mathcal{F}^2 = (7,18,9,40)$ violate the same conditions.
In particular, $\mathcal{F}(P^4)$ is not convex:

$$
\mathcal{F}(P^4) \neq \operatorname{conv} \mathcal{F}(P^4) \cap \mathbb{Z}^4.
$$

(It is also known that $\operatorname{conv} \mathcal{F}(P^4)$ is not closed.)

(4.3) For all $i < j$, the possible pairs (f_i, f_j) for 4-polytopes were determined by Grünbaum [10], Barnette [2] and Barnette & Reay [3], see [7]. We comment on three specific values:

- Barnette [2] proved $(f_1, f_2) \neq (18, 16)$. This value has to be added to the list of special "impossible pairs" in [2] and in [7].
- In Barnette's paper, no proof is given for $(f_1, f_2) \neq (26, 21)$. However, this can be derived from the inequalities in (3), see [11, page 48].
- [2] lists $(f_1, f_2) = (32, 25)$ as an impossible pair, but polytopes with these values exist.

The impossible pairs from the above characterizations allow one to exclude eleven integer vectors that satisfy the conditions of (3.0)–(3.2). These are the ones listed in (3.3). All these "excluded" flag-vectors satisfy $f_0 \leq 20$ and $f_3 \leq 20$.

(5) For our computational study, we have enumerated all the "potential flag-vectors" that satisfy the conditions of (3) in the range $f_0 \leq 20$ and $f_3 \leq 20$: There are exactly

$$n(f_0 \leq 20, f_3 \leq 20) \quad = \quad 54768$$

such potential flag-vectors.

In particular, note that $f_0 \leq 8$ implies $f_3 \leq 20$ by the Upper Bound Theorem. There are

4 potential flag-vectors for $f_0 = 6$,
20 potential flag-vectors for $f_0 = 7$, and
86 potential flag-vectors for $f_0 = 8$.

We managed to generate 8363 flag-vectors in the range $\max\{f_0, f_3\} \leq 20$ by combining the following methods:

- generation of random polytopes whose vertices have small integer coordinates,
- generation of all 19 flag-vectors of simplicial polytopes in the range, according to the g-theorem,
- generation of all 165 flag-vectors of pyramids and bipyramids over 3-polytopes,
- generation of all 4-dimensional 0/1-polytopes, which yields 139 different flag-vectors,
- generation of flag-vectors by "cutting off simple vertices" whenever existence of such a vertex was established, and
- dualization: with each flag-vector we included the flag-vector of the dual polytope.

A great part of the analysis, in particular, the generation and the analysis of random polytopes and of the 0/1-polytopes, was performed in the POLYMAKE system of Gawrilow & Joswig [9].

(6) For small numbers of vertices, we obtained actual polytopes for nearly all potential flag-vectors.

Theorem. *All potential flag-vectors (that satisfy the conditions of section (3)) with $f_0 \leq 7$ occur as flag-vectors of convex 4-polytopes, with the sole exception of*

$$(7, 17, 9, 39).$$

For $f_0 = 8$, convex polytopes exist for all potential flag-vectors, except for the following twelve:

$(8, 19, 9, 42)$	$(8, 20, 8, 40)$	$(8, 20, 9, 43)$
$(8, 20, 10, 46)$	$(8, 21, 8, 39)$	$(8, 21, 12, 51)$
$(8, 22, 13, 55)$	$(8, 23, 14, 59)$	$(8, 24, 15, 63)$
$(8, 21, 12, 52)$	$(8, 22, 8, 38)^*$	$(8, 22, 13, 56)$

Of these twelve, the first nine lie in the interior of the convex hull of actual flag-vectors of 4-polytopes, while the last three lie outside the convex hull.

Proof. The potential flag-vector $(7, 17, 9, 39)$ was found by Bayer [5]; it is impossible since it is not assumed in the complete classification of all 4-polytopes with 7 vertices [10, Sect. 6.2]. The non-realizability of the twelve vectors above follows from the classification of 4-polytopes with 8 vertices by Altshuler & Steinberg [1]. For this, we have computed the flag-vectors of all 4-polytopes with 8 vertices from the data of the combinatorial types that are given in Tables 1 and 3 of [1]. There are 1294 different combinatorial types, but they have only 74 different flag-vectors — the 12 "potential flag-vectors" listed above are non-realizable.

It is an interesting challenge to provide *new* flag-vector inequalities that would "cut off" some of the twelve potential flag-vectors given above. See [11, page 74] for more non-realized potential flag-vectors.   □

(7) There is still a long way to go until we can claim to "understand" the flag-vectors, or even just the f-vectors of 4-polytopes. For example, the "diagonal" case of polytopes with $f_0 = f_3$ shows our ignorance. This case is also relevant for the question about "fat-lattice" 4-polytopes, as asked by Avis, Bremner & Seidel [4]: how large is

$$f(n) \quad := \quad \max\{f_1 : (f_0, f_1, f_3, f_{03}) \in \mathcal{F}(\mathcal{P}^4), \ f_0, f_3 \leq n\}?$$

This question is of considerable importance for the complexity of convex-hull algorithms!

Here we note the following: addition of (3.1.1) and (3.2.3) yields

$$2f_1 \ \leq \ f_0 + f_3 + \binom{f_0}{2} \ = \ \frac{f_0^2 + f_0 + 2f_3}{2} \ \leq \ \frac{(\max\{f_0, f_3\})^2 + 3\max\{f_0, f_3\}}{2}$$

and thus

$$f_1 \ \leq \ \left\lfloor \frac{n^2 + 3n}{4} \right\rfloor \qquad \text{for } n := \max\{f_0, f_3\}.$$

And indeed, this inequality is sharp for *potential f-vectors*: for example, for $f_0 = f_3$ we have the potential f-vectors.

$(4k, 4k^2 + 3k, 4k, 4k^2 + 11k)$	for $k \geq 2$,
$(4k + 1, 4k^2 + 5k + 1, 4k + 1, 4k^2 + 13k + 3)$	for $k \geq 1$,

$$(4k + 2, 4k^2 + 7k + 2, 4k + 2, 4k^2 + 15k + 6) \quad \text{for } k \geq 2,$$
$$(4k + 2, 4k^2 + 7k + 2, 4k + 2, 4k^2 + 15k + 7) \quad \text{for } k \geq 1,$$
$$(4k + 3, 4k^2 + 9k + 4, 4k + 3, 4k^2 + 17k + 10) \quad \text{for } k \geq 2, \quad \text{and}$$
$$(4k + 3, 4k^2 + 9k + 4, 4k + 3, 4k^2 + 17k + 11) \quad \text{for } k \geq 1.$$

However, these vectors are *potential*: for $f_0 = 8$ ($k = 2$), the flag-vector $(8, 22, 8, 38)$ does **not** appear for any polytope (see the theorem above), and the maximal f_1 (assuming $f_0 = f_3$) is obtained by $(8, 21, 8, 38)$. Similarly, for $f_0 = 18$ the largest f_1 for a *potential* flag-vector is attained by $(18, 94, 18, 130)$, while the largest f_1 we actually *found* was in $(18, 61, 18, 102)$. So the gap (measured, e. g., in the ratio $\frac{f_1}{f_0}$), widens as $f_0 = f_3$ gets larger. And indeed, a result of Edelsbrunner & Sharir [8] implies that

$$f_{03} = O(f_0^{2/3} f_3^{2/3}),$$

which is $f_{03} = O(f_0^{4/3})$ in the "diagonal case" where $f_0 = f_3$. Unfortunately, it seems to be hard to get an explicit value for the constant hidden in the "O" of this result.

Here is a conjectured inequality for flag-vectors by M. Bayer [5, p. 149]:

$$2f_0 - f_1 + 3f_3 \geq 15.$$

We found no actual polytopes that would violate this conjecture. However, counter-examples can be found among the sequences of flag-vectors listed above, where f_1 grows quadratically with $n := f_0 = f_3$. As specific examples, we note the potential flag-vectors

$(13, 51, 13, 77)$	yields $2f_0 - f_1 + 3f_3 =$	$14,$
$(18, 91, 18, 127)$	yields $=$	$-1, \quad$ and
$(20, 115, 20, 155)$	yields $=$	$-15.$

Here is another conjectured inequality for flag-vectors [5, p. 145]:

$$2f_0 + f_1 + 3f_3 - f_{03} \geq 15.$$

Again we found no actual polytopes that violate this conjecture, but there are counterexamples among the potential f-vectors, such as

$(15, 41, 15, 102)$	yields $2f_0 + f_1 + 3f_3 - f_{03} =$	$14,$
$(20, 61, 20, 162)$	yields $=$	$-1, \quad$ and
$(20, 62, 20, 166)$	yields $=$	$-4.$

A much stronger inequality, in some sense the "strongest possible" one, which would imply all those just mentioned from [5] and from [8], was suggested by Billera and Ehrenborg (personal communication):

$$10f_0 - f_1 + 9f_3 - 2f_{03} \geq 45.$$

Again, our data give lots of "potential f-vectors" that violate this, but no real counter-example.

Acknowledgement. Thanks to Marge Bayer for helpful comments and encouragement.

References

[1] A. ALTSHULER & L. STEINBERG: *The complete enumeration of the 4-polytopes and the 3-spheres with eight vertices, Pacific J. Math.* **117** (1985), 1–16.

[2] D. W. BARNETTE: *The projection of the f-vectors of 4-polytopes onto the (E, S)-plane, Discrete Math.* **10** (1974), 201–216.

[3] D. W. BARNETTE & J. R. REAY: *Projections of f-vectors of four-polytopes, J. Combinatorial Theory*, Ser. A **15** (1973), 200–209.

[4] D. AVIS, D. BREMNER & R. SEIDEL: *How good are convex hull algorithms? Computational Geometry: Theory and Applications* **7** (1997), 265–301.

[5] M. M. BAYER: *The extended f-vectors of 4-polytopes, J. Combinatorial Theory*, Ser. A **44** (1987), 141–151.

[6] M. M. BAYER & L. J. BILLERA: *Generalized Dehn-Sommerville relations for polytopes, spheres and Eulerian partially ordered sets, Inventiones Math.* **79** (1985), 143–157.

[7] M. M. BAYER & C. W. LEE: *Combinatorial aspects of convex polytopes*, in: "Handbook of Convex Geometry" (P. Gruber and J. Wills, eds.), North-Holland, Amsterdam 1993, pp. 485–534.

[8] H. EDELSBRUNNER & M. SHARIR: *A hyperplane incidence problem with applications to counting distances,* in: "Applied Geometry and Discrete Mathematics – The Victor Klee Festschrift" (P. Gritzmann and B. Sturmfels, eds.), DIMACS Series in Discrete Mathematics and Theoretical Computer Science **4**, Amer. Math. Soc. 1991, pp. 253–263.

[9] E. GAWRILOW & M. JOSWIG: *POLYMAKE – a framework for analyzing convex polytopes*, this volume, 43–73.

[10] B. GRÜNBAUM: *Convex Polytopes*, Interscience, London 1967.

[11] A. HÖPPNER: *F-Vektoren und Fahnenvektoren von 4-dimensionalen Polytopen*, Diplomarbeit, TU Berlin 1998.

[12] E. STEINITZ: *Über die Eulerschen Polyederrelationen, Archiv für Mathematik und Physik* **11** (1906), 86–88.

Andrea Höppner
Current address:
Andrea Tan
Hedderichstr. 40
D-60594 Frankfurt/M., Germany
Andrea.Tan@tlc.de

Günter M. Ziegler
Dept. Mathematics, MA 7-1
Technische Universität Berlin
Str. des 17. Juni 136, 10623 Berlin, Germany
[hoeppner,ziegler]@math.tu-berlin.de

Extremal Properties of 0/1-Polytopes of Dimension 5

Oswin Aichholzer*

Abstract. In this paper we consider polytopes whose vertex coordinates are 0 or 1, so called 0/1-polytopes. For the first time we give a complete enumeration of all 0/1-polytopes of dimension 5, which enables us to investigate various of their combinatorial extremal properties.

For example we show that the maximum number of facets of a five-dimensional 0/1-polytope is 40, answering an open question of Ziegler [25]. Based on the complete enumeration for dimension 5 we obtain new results for 2-neighbourly 0/1-polytopes for higher dimensions.

1. Introduction

Among the simplest high-dimensional geometric objects is the d-dimensional hypercube (d-cube). It is a convex polytope that, when having unit edge length, may be expressed as $\mathcal{C}^d = [0,1]^d$.

Despite of its simple definition, $\mathcal{C}^d$ has been an object of study from various different points of view. Purely combinatorial properties of $\mathcal{C}^d$, mainly involving certain subgraphs formed by its edges and vertices (the latter are just the various d-tuples of binary digits) have been investigated extensively in coding theory, communication theory, learning theory, and many other fields; see, e.g., [4, 12, 21]. This reveals the fact that the hypercube is the most natural geometric structure combining all binary digits of length d.

The theory of convex polytopes provides classical results concerning sections and projections of hypercubes (see e.g. [8, 13]) as well as for so-called subpolytopes of the d-cube [22]. In this paper we consider the convex hull of a subset $V \subseteq \{0,1\}^d$ of hypercube vertices. A polytope P representing the convex hull of V is called a *0/1-polytope*. Thus the facets of 0/1-polytopes are supported by hyperplanes spanned by d-cube vertices. This set is investigated e.g. in [2]. For a nice introduction to 0/1-polytopes see Ziegler [23].

Motivation for considering 0/1-polytopes comes from Combinatorial Optimization (e.g. the Traveling Salesman Polytope, or the Assignment Polytope) and many other fields, see e.g. Ziegler [24, 23, 25].

* Research supported by the Spezialforschungsbereich F 003, *Optimierung und Kontrolle*.

A point $p \in P$ is an *extreme point* of P provided $x, y \in P$, $0 < \lambda < 1$, and $p = \lambda x + (1 - \lambda)y$ imply $p = x = y$. In other words, p is an extreme point of P if it does not belong to the relative interior of any segment contained in the closure of P. It is not hard to see that any vertex of V is an extreme point of P: consider the d-dimensional hypersphere S^d with center $(\frac{1}{2}, \dots, \frac{1}{2})$ and radius $\frac{\sqrt{d}}{2}$. All vertices of the d-cube C^d lie on S^d and thus any subset V of vertices of C^d entirely consists of extreme points. Thus a 0/1-polytope P is a so-called subpolytope of the hypercube.

In Section 2 we give for the first time a complete enumeration of all 0/1-polytopes of dimension 5. This enables us to investigate various combinatorial extremal properties, see Section 3. One main result we obtain from this enumeration is to determine the maximum number of facets of a five-dimensional 0/1-polytope. We show that this maximum is 40, answering an open question of Ziegler [25]. We also give extremal results on the edge skeleton of five-dimensional 0/1-polytopes, e.g. we give an example with the maximal number of 112 edges. Moreover we show that the degree of a vertex can exceed 2^{d-1} already for $d \geq 5$.

Based on the complete enumeration for dimension 5 we obtain as a spin-off new results for 2-neighbourly 0/1-polytopes for higher dimensions, see Section 4.1.

2. Complete enumeration for dimension 5

Figure 1 shows all types of three-dimensional 0/1-polytopes, along with the number of vertices, edges, facets, and the number of geometrically equivalent copies, respectively. We call two polytopes 0/1-equivalent if they can be transformed into each other by a sequence of rotations or reflections, see Section 2.1 for more details. Thus for $d = 3$ there are 12 different types of 0/1-polytopes, each representing an equivalence class of the given cardinality. In total this gives 151 0/1-polytopes in three-space. The remaining 105 possible subsets of vertices of the 3-cube span polytopes of dimension at most two.

Note, however, that we consider 0/1-equivalence classes as opposed to combinatorially equivalent polytopes as defined in [22]. From Figure 1 it is easy to see that there exist only 8 combinatorially different 0/1-polytopes for dimension three. See also [23] for different types of equivalence between 0/1-polytopes and their relationship.

For arbitrary dimension d there are $2^{(2^d)}$ possible vertex subsets of the d-cube. Most of them span a d-dimensional 0/1-polytope. In fact, it has been shown recently that there are at least $2^{(2^{d-2})}$ different combinatorial types of d-dimensional 0/1-polytopes for $d > 5$, see revised printing of [22]. For $d \leq 4$ a complete enumeration of all sets can simply be done by using standard software. But starting with dimension 5 the number of possible sets increases tremendously. For $d = 6$ there are already about $1.8 \cdot 10^{19}$ possible sets and at least $2^{(2^6)}/(6! \, 2^6) \approx 4.0 \cdot 10^{14}$ symmetry classes, i.e. dimension 6 is even infeasible with use of symmetry. Thus dimension 5 seems to be the first non-trivial and last feasible dimension for enumerating all

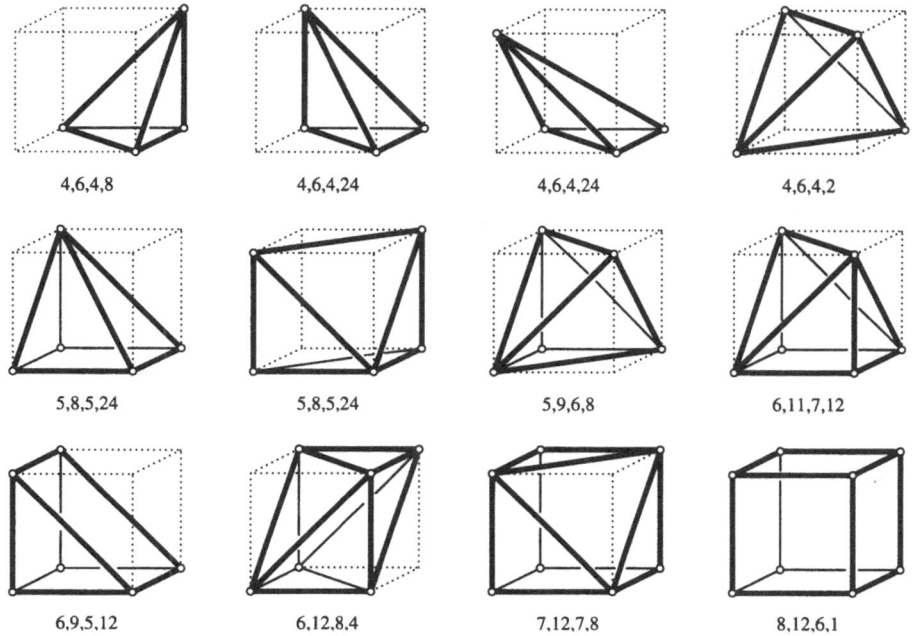

FIGURE 1. Three-dimensional 0/1-polytopes.

such sets. In this section we give for the first time a complete enumeration of all non-equivalent 0/1-polytopes for dimension 5. To this end we use some results of [2] and several observations given below.

A straightforward method to enumerate all subsets of vertices of $\mathcal{C}^5$ would be impractical because of the huge number of $2^{(2^5)} \approx 4.3 \cdot 10^9$ such sets. Thus if we could compute about one hundred 5d-convex hulls per second, it would take over one year to enumerate all sets. Fortunately many of these sets are equivalent, i.e. are reflections and rotations of each other. Thus for every equivalence class we need to consider only one representative.

We therefore divide the enumeration into two parts. First we generate exactly one representative vertex set for every equivalence class. We then can easily perform various investigations of these sets in reasonable time, e.g. hunting for the 0/1-polytope with the maximal number of facets.

2.1. Generating representatives of equivalence classes

Let us first consider the equivalence relations. A *reflection* of a vertex set V of the d-cube for an axis i exchanges all zeros and ones on the ith bit position of all vertices of V. Reflections at k axes are thus reflections at a $(d - k)$-subspace. By combining reflections for a certain set of axes we can thus transform a given vertex of V to an arbitrary vertex of the hypercube. For example one can "move" an arbitrary vertex of the set to the origin. Clearly reflections do only change the orientation of the vertex set, but not the properties of their convex hull.

The second operation is called *rotation*. Here we allow arbitrary permutations of the coordinate axes. This operation reflects the fact that we do not weigh the bits of a vertex according to their position, i.e. all directions in d-space have the same significance. Again, rotations might affect the absolute position of the vertices and the orientation of the set, but the structure of their convex hull remains unchanged.

A *class of vertex sets* contains all vertex sets which can be transformed into each other by repeated application of the above two operations. The polytopes represented by these sets are *0/1-equivalent*. For simplification we will call them equivalent in the remaining paper, but see [23] for alternative equivalence relations. Obviously all polytopes within a class have geometrically equivalent convex hulls and we need only to consider one vertex set per class.

Each vertex of the d-cube can be associated, in a natural way, with its *value*, the decimal value of the binary string corresponding to the vector coordinates. The *vertex vector* $b = (b_0, b_1, \dots, b_{2^d-1})$ of a set V of vertices from the d-cube is a binary vector of length 2^d with $b_i = 1$ if and only if the vertex with value i is an element of V. For example the vertex vector of the set $\{(0,0,0),(0,1,1),(1,0,1)\}$ is $b = (1,0,0,1,0,1,0,0)$. Clearly, reflections and rotations of a vertex set change its vertex vector.

For any class of vertex sets we can select the set with the lexicographically maximal vertex vector as its representative. In the following we will show that these sets have properties which can be utilized to generate them.

Consider a vertex set V with n vertices of the d-cube and suppose that the vertices are in increasing lexicographical order. We then build a binary $(n \times d)$-matrix $A(V) = (v_1, \dots, v_n)^T$ with the vertices of V as its rows. A *block* of $A(V)$ is a maximal set of indices of consecutive identical column vectors. This means that permuting the columns within a block does not change the matrix $A(V)$ and thus does not affect the vertices of V.

A vertex $v_i \in V$ has the *block property* if for any block b built by $A(v_1, \dots, v_{i-1})$ and for any pair $k < l$ of indices within b we have $v_i(k) \leq v_i(l)$, where $v_i(k)$ denotes the kth bit of v_i. In other words, within a block built by the previous vertices, v_i always has the form $(0, 0, \dots, 0, 1, 1, \dots, 1)$, i.e. "all ones are shifted to the right".

A set V of vertices satisfies the block property if every vertex of V has the block property. If this is the case it follows by induction that identical column vectors of $A(V)$ are consecutive (recall that the vertices in V have to be in increasing lexicographical order).

Let us give an example for C^9 with $n = 4$:

$$
\begin{aligned}
v_1 &= (0,0,0,0,0,0|1,1,1) \\
v_2 &= (0,0|1,1,1,1|0,0,0) \\
v_3 &= (0|1|0,0,0|1|1,1,1) \\
v_4 &= (1|1|0|1,1|0|0|1,1)
\end{aligned}
$$

$A(v_1, v_2, v_3)$ has blocks of length 1,1,3,1 and 3, respectively. Thus the additional vertex v_4 satisfies the block property. This also holds for v_1, v_2 and v_3, so that the whole vertex set $V = \{v_1, v_2, v_3, v_4\}$ has the block property.

Lemma 1. *In a class of equivalent vertex sets, a lexicographical maximal vertex vector is attained by a vertex set which (a) contains the origin and (b) has the block property.*

Proof. We have to show that any vertex set $V = \{v_1, \ldots, v_n\}$ which does not have properties (a) and (b) can be transformed into an equivalent set V' with a (lexicographically) larger vertex vector, satisfying (a) and (b).

First recall that if the set does not include the origin then by reflection we can move an arbitrary vertex of the set to the origin. This clearly increases the vertex vector. Now consider the vertices in increasing order. Suppose that a vertex v_i, $2 \le i \le n$, does not have the block property according to the vertices $v_1, \ldots, v_{i-1}$. By permutation of the positions within the blocks we can rearrange the bits of v_i such that it satisfies the block property. This reduces the value of v_i and clearly affects none of the vertices $v_1, \ldots, v_{i-1}$. Note that although the permutations might change $v_{i+1}, \ldots, v_n$, too, the reduction of the value of v_i still dominates the lexicographic order. Thus v_i does not only have the block property, but we also increased the vertex vector of the set.

By induction we conclude that a vertex set attaining the maximal vertex vector has to satisfy properties (a) and (b). □

Note, however, that in an equivalence class of vertices there might be more than one vertex set with properties (a) and (b). Therefore the algorithm to generate exactly one representative vertex set per equivalence class has two stages.

In the first part we generate only vertices which satisfy properties (a) and (b) of Lemma 1. We use a set of nested for-loops with fixed first element $v_0 = (0, \ldots, 0)$. For every vertex v_i, $0 \le i \le 2^5 - 1$, we precompute two bit masks. The first mask has a '1' at every bit position where v_i forces a new block to start, i.e. for the second bit of every '01' or '10' combination of v_i. When adding a new vertex to a set we can therefore update the block information by a simple constant time binary OR operation (assuming constant dimension d). The second mask has a '1' for the second bit of every '10' pair of v_i. In order to ensure that an additional vertex v_i has the block property there must start a new block of the previous vertex set for every '1' in this mask. This means that for every '1' in the second mask of v_i there must be a '1' in the block mask of the previous vertices, which is just the OR-combination of their first masks. For constant dimension we can thus decide in constant time whether a new vertex satisfies the block property, i.e. whether it has to be considered. An example of a vertex (for dimension 10) and its two bit masks is given below.

$$
\begin{aligned}
v_i &= \quad (0,0,1,1,0,0,1,0,1,1) \\
mask_1 &= \quad (0,0,1,0,1,0,1,1,1,0) \\
mask_2 &= \quad (0,0,0,0,1,0,0,1,0,0)
\end{aligned}
$$

k	$\binom{2^5}{k}$	generated sets	equivalence classes	class cardinality min. / average / max.		
0	1	1	1	1	1.0	1
1	32	1	1	32	32.0	32
2	496	5	5	16	99.2	160
3	4960	30	10	160	496.0	960
4	35960	230	47	40	765.1	1920
5	201376	1821	131	32	1537.2	3840
6	906192	12789	472	32	1919.9	3840
7	3365856	73880	1326	32	2538.4	3840
8	10518300	344054	3779	20	2783.4	3840
9	28048800	1300574	9013	160	3112.0	3840
10	64512240	4053531	19963	16	3231.6	3840
11	129024480	10582906	38073	32	3388.9	3840
12	225792840	23453731	65664	16	3438.6	3840
13	347373600	44578853	98804	160	3515.8	3840
14	471435600	73233239	133576	80	3529.3	3840
15	565722720	104566543	158658	32	3565.7	3840
16	601080390	130282523	169112	2	3554.3	3840
Σ	2448023843	392484711	698635			

TABLE 1. Number and cardinality of equivalence classes of vertex sets of the 5-cube.

With these two masks we can generate all vertex sets with properties (a) and (b) of Lemma 1 very efficiently.

For the second part of our algorithm we apply all possible reflections and rotations to the generated sets. For dimension 5 this gives up to $2^5 \cdot 5! = 3840$ possible copies per set. A vertex set is stored only if the corresponding vertex vector yields the maximum among all resulting vertex vectors. In this way we guarantee that for every equivalence class we store one unique representative.

Table 1 lists the results for $k = 0 \ldots 16$ vertices of $\mathcal{C}^5$. Sets with $k > 16$ vertices can be seen as the complement of sets with $32 - k$ vertices and can easily be obtained from them. The first three columns give the number of all vertex sets, the number of generated sets (part one of the above description) and the number of equivalence classes, respectively. It can be seen that less than $\frac{1}{6}$ of all vertex sets with ≤ 16 vertices satisfy properties (a) and (b) of Lemma 1. Thus we needed to apply the more time consuming rotations and reflections only to this smaller set. The third column shows that there are less than 700000 equivalence classes for $k \leq 16$ and about 525000 for $k > 16$. Therefore the subsequent investigations have to enumerate about 1.2 million sets, which is less than 0.03% of the initial number of sets. This enables us to consider various related questions within a couple of hours instead of month or years.

The number of equivalence classes can also be computed by Pólya's theory of counting [20, 5]. Based on Burnside's lemma [7] the cycle index of the permutation group is built, which then, by Pólya's theorem, gives the number of equivalence classes, see e.g. [15, 19] for more details. The results obtained from this theoretical approach coincide with the output of our program. However, with Pólya's counting theorem it is not possible to obtain a representative vertex set for each equivalence class.

The last three columns of Table 1 show the minimum, average and maximal cardinality of the equivalence classes. There are highly symmetric vertex sets, e.g. for $k = 16$ there is a class with only two sets: The first consists of all vertices with an even number of ones and the second is its complement. On the other hand there are classes with the maximal number of 3840 different sets. On average a class contains over 3000 sets, which results in the comparatively small number of classes.

We used 4 bytes to store the 32-bit vertex vector of every representative vertex set and one additional byte to store the number of vertices, which is clearly redundant but useful for fast access. This results in about 3.5 MB of data, which can easily be handled by any standard computer equipment. In fact we used a simple Personal Computer to compute these results. In a second file we also stored the cardinality of each class (about 1.4 MB). To enable the interested reader to carry out her or his own investigations the files are available upon request from the author.

2.2. Computing the convex hull of 0/1-polytopes

The clear advantage of our approach is that we need to precompute the set of all representative vertex sets only once. Any task which requires a complete enumeration of all subpolytopes of the 5-cube can then be computed easily within a few hours on a standard computer equipment. In this subsection we give a brief description of some convex hull algorithms to derive combinatorial properties of 0/1-polytopes in dimension 5. For further investigations we consider the number of facets and the number of edges as well as properties like 2-neighbourliness, simpliciality and simplicity. In other words we concentrate on the 1-dimensional and $(d - 1)$-dimensional faces of these 0/1-polytopes. By complete enumeration of the convex hulls of all representative vertex sets generated in the last subsection we thus get all extremal results on edges and facets for dimension 5.

First we ran a simple implementation of an incremental method to compute higher dimensional convex hulls for point sets in $\mathbb{R}^d$. The algorithms worked well and within several hours we could enumerate all classes. Note that when using a standard convex hull algorithm one has to be careful, since vertices of $\mathcal{C}^d$ are highly degenerate in the sense that there are many linear dependences between them. This may cause numerical troubles in computing the convex hull and in fact many of the existing software packages fail on this task (compare e.g. [6]). For example perturbation methods may result in superpolynomially sized convex hulls for the d-cube [3].

For the second approach we used a tailor-made algorithm for C^5. We utilized the fact that we have exactly 3254 hyperplanes spanned by vertices of the 5-cube, see [2]. This means that for 0/1-polytopes there are 3254 possible ways to support their facets. For every hyperplane we precompute three vertex vectors. The first two contain all vertices of C^5 which are in or above (in or below, respectively) the hyperplane. The third set contains all vertices covered by the hyperplane. Using the first two vertex vectors we can by simple bit operations decide in constant time, whether a hyperplane supports a given set V of vertices or not. Using the third vertex vector we can identify all vertices of V which are covered by the hyperplane. Solving a simple linear system of equations shows us whether these vertices span the hyperplane. If and only if this is the case the hyperplane counts for one facet. Reiterating this for all 3254 possible hyperplanes gives the number of facets of the convex hull of V.

The second approach worked quite well. On a standard Personal Computer (166 MHz Pentium$^©$ processor) we needed less than one hour to examine all classes of up to 16 vertices, and an additional hour for the complementary sets.

To compute the number of edges of the convex hull of a vertex set we again used both a standard technique and a tailor-made algorithm. The later version is based on Observation 1 below. We say a subcube C^k, $k \le d$, is *spanned* by two vertices v_i and v_j of the d-cube, if the edge $v_i v_j$ is a diagonal of C^k. Thus the dimension of C^k is the Hamming distance between v_i and v_j. Note that C^k need not include the origin (of C^d) in this case. Let now V be a subset of the vertices of C^d and $V' \subseteq V$ a subset of C^k. Since the edges of the convex hull of a k-dimensional point set do not change when we add points in higher dimensions only on one side of this set, it follows that all edges of the convex hull of V' are part of the convex hull of V. We thus get the following observation.

Observation 1. *Let V be a vertex set of the d-cube. An edge $e = v_i v_j$, $v_i, v_j \in V$, is part of the convex hull of V iff e is part of the convex hull of the subset induced by the subcube spanned by v_i and v_j.*

Let V be a subset of vertices of the d-cube. From Observation 1 it follows that we can compute the edges of the convex hull of V by inserting all edges spanned by vertices of V in order of non-decreasing length. For every edge e we consider the subcube C^k spanned by its two endpoints and the subset $V' \subseteq V$ induced by C^k. If the edge skeleton of V' has more than one main diagonal (with respect to C^k) then none of them can belong to the convex hull of V' and thus V. Else we have to check whether the diagonal e is part of it. To this end consider the set E of edges of the (partially) convex hull of V' emanating from one endpoint of e. Note that all these edges are shorter than e and therefore we already know that they are part of the convex hull. Then e belongs to the convex hull iff e is no positive linear combination of edges from E. Thus we have to solve a corresponding linear-programming problem for E and e is part of the convex hull iff the system has no feasible solution. Using this approach we computed the number of edges and the maximal vertex degree of all 0/1-polytopes of dimension 5.

Based on the database obtained by the above investigations we have implemented a small program which enables us to generate the set of all d-dimensional 0/1-polytopes, $2 \leq d \leq 5$, with prescribed properties within a few seconds.* For example one can ask for all five-dimensional 0/1-polytopes with at least 10 vertices and more than 30 facets that are simplicial.[†] Moreover the program offers the option to generate the output in POLYMAKE-format (see [10]), such that further investigations or the visualization of the resulting polytopes can easily be done by this nice tool.

3. Extremal properties of 0/1-polytopes of dimension 5

In this section we summarize the results obtained from the classification of 0/1-polytopes described in the last section.

3.1. Counting facets

Motivation for counting the number of facets of 0/1-polytopes is given by Ziegler [24, 25], Henk et.al. [14] and Kortenkamp et.al. [18]. We denote the maximal number of facets of a d-dimensional 0/1-polytope with $f(d)$. It is well known and easy to check that $f(d) = 2^d$ for $d \leq 4$. For dimension 5 a 0/1-polytope with 40 facets is known, i.e. $f(5) \geq 40$, but it was unknown whether this bound is tight. The general question is how fast $f(d)$ grows with d. In particular it is unknown if there exists a constant c such that $f(d) < c^d$ for all d [24, 25]. The best known upper bounds can be found in Fleiner et.al. [9]. The latest results in the race for 0/1-polytopes with many facets of up to dimension 13 can be obtained from the homepage of Kortenkamp [17] and the polytopes are available from the polymake database [10, 11].

Table 2 summarizes the results for dimension 5. The first column gives the number k of vertices and the next two columns give the number of vertex sets and the number of classes, respectively, which provide a full-dimensional convex hull. In other words this is the number of sets (classes) where not all vertices lie in a hyperplane. For counting the number of facets or edges we only took these sets into account. Columns 4 to 6 give the minimal, average and maximal number of facets for a fixed number of vertices. Figure 2 shows the corresponding curves. Note that the average is taken over the full-dimensional vertex sets, rather than over the classes. For completeness we give similar results for the 4-cube in Table 3.

---

*This program can also be accessed via e-mail: write to cggg@igi.tu-graz.ac.at with the body "01poly" for further information. In addition the database includes all six-dimensional 0/1-polytopes with up to 12 vertices (94826705 classes).
[†] For the interested reader: There are 7 equivalence classes satisfying these properties, giving together 3008 such 0/1-polytopes, with 32 facets each.

k	full-dim. vertex sets	full-dim. classes	number of facets min. / average / max.		
6	556192	237	6	6.00	6
7	2925056	1062	7	8.35	12
8	10071580	3462	7	11.06	20
9	27678880	8781	7	13.84	26
10	64264048	19767	7	16.41	32
11	128892000	37976	8	18.56	34
12	225738120	65600	7	20.15	40
13	347356800	98786	8	21.16	38
14	471432000	133565	8	21.67	38
15	565722240	158656	8	21.80	36
16	601080360	169110	8	21.64	35
17	565722720	158658	9	21.30	34
18	471435600	133576	8	20.82	31
19	347373600	98804	10	20.24	30
20	225792840	65664	9	19.57	28
21	129024480	38073	10	18.83	27
22	64512240	19963	10	18.03	26
23	28048800	9013	10	17.19	22
24	10518300	3779	9	16.35	20
25	3365856	1326	12	15.52	18
26	906192	472	11	14.74	16
27	201376	131	12	13.99	15
28	35960	47	11	13.28	14
29	4960	10	12	12.58	13
30	496	5	11	11.84	12
31	32	1	11	11.00	11
32	1	1	10	10.00	10
Σ	4292660729	1226525			

TABLE 2. 5-cube: number of full-dimensional vertex sets, equivalence classes and number of facets of 0/1-polytopes with k vertices.

Table 2 shows that the average number of facets is maximized for $k = 15$, whereas the "best" single example has fewer vertices, namely $k = 12$. This "shift" can also be observed for $d = 3$ and $d = 4$, cf. Figure 1 and Table 3. Thus it might also be true for higher dimensions and has therefore to be taken into account when looking for "maximizing" sets by investigation of random vertex sets. The most promising vertex sets for maximizing the number of facets might have significantly less vertices than the sets with the maximal average number of vertices.

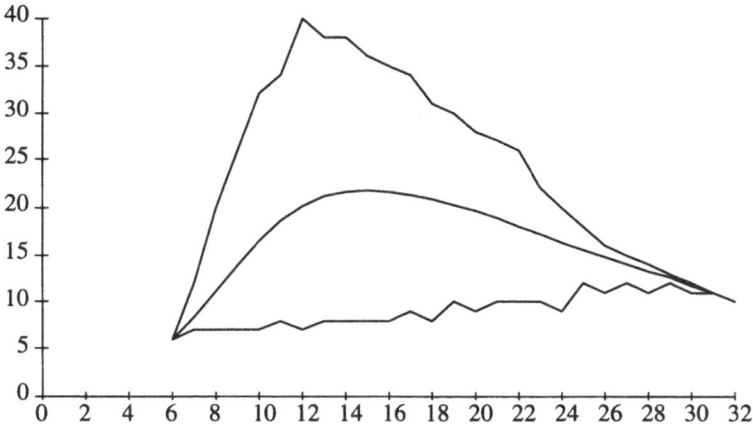

FIGURE 2. Minimal, average and maximal number of facets of
five-dimensional 0/1-polytopes over the number of vertices.

k	full-dim. vertex sets	full-dim. classes	number of facets min. / average / max.		
5	3008	17	5	5.00	5
6	7408	40	6	6.59	9
7	11280	54	6	8.29	13
8	12850	72	6	9.84	16
9	11440	56	6	10.98	15
10	8008	50	7	11.52	14
11	4368	27	8	11.46	13
12	1820	19	7	10.99	12
13	560	6	10	10.37	11
14	120	4	9	9.73	10
15	16	1	9	9.00	9
16	1	1	8	8.00	8
Σ	60879	347			

TABLE 3. 4-cube: number of full-dimensional vertex sets, equivalence classes and number of facets of 0/1-polytopes with k vertices.

Our results show that there exists only one unique vertex set of $\mathcal{C}^5$ (up to
hypercube symmetries) with 12 vertices and 40 facets and that this is indeed the
maximal number of facets for dimension 5.

$$
\begin{array}{lll}
v_1 = (0,0,0,0,0) & v_5 = (0,1,1,0,1) & v_9 = (1,0,1,1,1) \\
v_2 = (0,0,0,0,1) & v_6 = (0,1,1,1,1) & v_{10} = (1,1,0,0,1) \\
v_3 = (0,0,1,1,0) & v_7 = (1,0,0,1,1) & v_{11} = (1,1,0,1,0) \\
v_4 = (0,1,0,1,0) & v_8 = (1,0,1,0,0) & v_{12} = (1,1,1,0,0)
\end{array}
$$

TABLE 4. The 12 vertices of the five-dimensional 0/1-polytope with the maximal number of 40 facets.

Lemma 2. *The maximal number of facets of a five-dimensional 0/1-polytope is $f(5) = 40$. This number is achieved by the vertex set given in Table 4. It is unique up to hypercube symmetries.*

It is interesting to observe that the equivalence class of the "best" example given in Table 4 contains 192 vertex sets. Its facets are 30 simplices and 10 bipyramids. (All other sets with 12 vertices have ≤ 36 facets.)

Let us close this subsection with a remark on the vertex-facet ratio. Obviously a facet of a d-dimensional 0/1-polytope may contain up to 2^{d-1} vertices, half of the vertices of the d-cube. The other way round it turned out that for $d = 5$ one vertex might be contained in up to 24 facets (achieved e.g. by a set of 14 vertices with 57 edges and 29 facets).

3.2. Counting edges

Let us now consider the number of edges of 0/1-polytopes. Motivation stems from graph-based algorithms which use the fact that every vertex of a subset of hypercube vertices is extreme, i.e. lies on the convex hull. Thus for many problems local optimization yields a global optimum. Clearly the number of edges and the degrees of the vertices will affect the efficiency of the algorithms.

A trivial lower bound for the maximal number of edges is given by $d2^{d-1}$, the number of edges of the d-cube. Table 5 and Figure 3 show the minimal, average and maximal number of edges of 0/1-polytopes of dimension 5. Table 5 also gives the maximal degree of a vertex.

The maximal number of 112 edges is achieved by a single equivalence class of 60 0/1-polytopes with 24 vertices and 18 facets. The representative set of this class is given in Table 6.

But also for smaller vertex sets the number of edges might be rather large. For example for 10 vertices there still exists a 2-neighbourly polytope (see Section 4.1), which means that any pair of vertices is adjacent. Thus for small vertex sets the maximal number of edges increases near quadratically with the number of vertices. This can also be seen from Figure 4 where the average number of edges for 10-dimensional 0/1-polytopes is given. Figure 4 also gives the upper and lower bounds attained by 20 random vertex sets for each set size. We thus conclude that if the dimension is very high, e.g. $d \geq 100$, the number of edges of the convex hull of a

k	number of edges			max. degree
	min. /	average /	max.	
6	15	15.00	15	5
7	19	19.80	21	6
8	22	25.07	28	7
9	24	30.66	36	8
10	25	36.42	45	9
11	31	42.22	51	10
12	30	47.94	60	11
13	36	53.51	67	12
14	37	58.86	72	13
15	39	63.93	77	14
16	40	68.69	84	15
17	48	73.10	89	16
18	45	77.15	91	17
19	55	80.79	96	17
20	52	84.01	100	17
21	57	86.79	105	17
22	61	89.09	110	17
23	64	90.93	109	16
24	60	92.28	112	16
25	75	93.14	109	16
26	75	93.49	106	15
27	80	93.30	103	14
28	76	92.49	100	13
29	85	90.96	95	12
30	83	88.54	90	10
31	85	85.00	85	8
32	80	80.00	80	5

TABLE 5. Number of edges and maximal degree of five-dimensional 0/1-polytopes with k vertices.

random vertex set of the d-cube is likely to be quadratic in the number of vertices, see also Section 4.2.

Let us now consider the degree of the vertices. Clearly the average degree is just one half of the average number of edges. Thus we ask here only for the maximal degree a vertex can achieve. A lower bound of 2^{d-1} is given by the pyramid over C^{d-1}: Take one half of the cube, say all vertices on the "left" side, and one additional vertex on the "right" side. Then obviously the vertex on the right side is adjacent to all other vertices, i.e. has degree 2^{d-1}.

It suggests itself to conjecture that the above lower bound is tight. By complete enumeration it is easy to verify that this is in fact the case for $d \leq 4$. But surprisingly the conjecture does not hold in general: Table 7 gives an example of

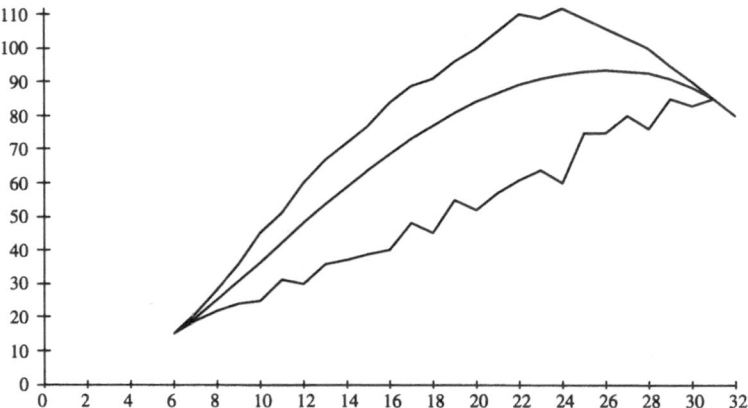

FIGURE 3. Minimal, average and maximal number of edges of
five-dimensional 0/1-polytopes over the number of vertices.

$v_1 = (0,0,0,0,1)$ $v_7 = (0,1,0,0,0)$ $v_{13} = (1,0,0,0,0)$ $v_{19} = (1,1,0,0,0)$
$v_2 = (0,0,0,1,0)$ $v_8 = (0,1,0,0,1)$ $v_{14} = (1,0,0,0,1)$ $v_{20} = (1,1,0,1,1)$
$v_3 = (0,0,1,0,0)$ $v_9 = (0,1,0,1,0)$ $v_{15} = (1,0,0,1,0)$ $v_{21} = (1,1,1,0,0)$
$v_4 = (0,0,1,0,1)$ $v_{10} = (0,1,0,1,1)$ $v_{16} = (1,0,0,1,1)$ $v_{22} = (1,1,1,0,1)$
$v_5 = (0,0,1,1,0)$ $v_{11} = (0,1,1,0,1)$ $v_{17} = (1,0,1,0,0)$ $v_{23} = (1,1,1,1,0)$
$v_6 = (0,0,1,1,1)$ $v_{12} = (0,1,1,1,0)$ $v_{18} = (1,0,1,1,1)$ $v_{24} = (1,1,1,1,1)$

TABLE 6. The 24 vertices of the five-dimensional 0/1-polytope
with the maximal number of 112 edges.

18 vertices of dimension 5, their degree and incident vertices. Here vertex v_{11} is
adjacent to all other vertices of the set, i.e. has degree 17.

3.3. Simple and simplicial 0/1-polytopes

A d-dimensional polytope is *simplicial* if all its proper faces are simplices, that is,
every facet has the minimal number of d vertices. It is *simple* if every vertex is
contained in the minimal number of only d facets (see e.g. [22]).

For example the hypercube is a simple 0/1-polytope and the convex hull of
any subset of vertices from $\mathcal{C}^d$ in general position gives a simplicial 0/1-polytope.

For dimension 5 there are 506 different classes of simplicial 0/1-polytopes,
giving together 1122976 such polytopes. Considering simple 0/1-polytopes we get
265 classes (with 574413 polytopes) of which 237 classes represent 5-simplices.
The remaining 28 classes build 6 combinatorial equivalent groups, all of whose are
the (combinatorial) product of simplices. This shows that all simple 0/1-polytopes
of dimension 5 are a product of simplices. Thus for simple 0/1-polytopes the 5-
cube is extremal: There is no other simple 0/1-polytope with more vertices, edges

vertices	degree	incident vertices
$v_1 = (0,0,0,0,0)$	8	$v_2, v_3, v_5, v_8, \ldots, v_{12}$
$v_2 = (0,0,0,0,1)$	7	$v_1, v_4, v_6, v_8, v_9, v_{11}, v_{13}$
$v_3 = (0,0,0,1,0)$	7	$v_1, v_4, v_7, v_8, v_{10}, v_{11}, v_{14}$
$v_4 = (0,0,0,1,1)$	7	$v_2, v_3, v_6, \ldots, v_8, v_{11}, v_{15}$
$v_5 = (0,0,1,0,0)$	7	$v_1, v_6, v_7, v_9, \ldots, v_{11}, v_{16}$
$v_6 = (0,0,1,0,1)$	7	$v_2, v_4, v_5, v_7, v_9, v_{11}, v_{17}$
$v_7 = (0,0,1,1,0)$	7	$v_3, \ldots, v_6, v_{10}, v_{11}, v_{18}$
$v_8 = (0,1,0,1,1)$	11	$v_1, \ldots, v_4, v_9, \ldots, v_{15}$
$v_9 = (0,1,1,0,1)$	11	$v_1, v_2, v_5, v_6, v_8, v_{10}, \ldots, v_{13}, v_{16}, v_{17}$
$v_{10} = (0,1,1,1,0)$	11	$v_1, v_3, v_5, v_7, \ldots, v_9, v_{11}, v_{12}, v_{14}, v_{16}, v_{18}$
$v_{11} = (1,0,1,1,1)$	17	$v_1, \ldots, v_{10}, v_{12}, \ldots, v_{18}$
$v_{12} = (1,1,0,0,0)$	8	$v_1, v_8, \ldots, v_{11}, v_{13}, v_{14}, v_{16}$
$v_{13} = (1,1,0,0,1)$	7	$v_2, v_8, v_9, v_{11}, v_{12}, v_{15}, v_{17}$
$v_{14} = (1,1,0,1,0)$	7	$v_3, v_8, v_{10}, \ldots, v_{12}, v_{15}, v_{18}$
$v_{15} = (1,1,0,1,1)$	7	$v_4, v_8, v_{11}, v_{13}, v_{14}, v_{17}, v_{18}$
$v_{16} = (1,1,1,0,0)$	7	$v_5, v_9, \ldots, v_{12}, v_{17}, v_{18}$
$v_{17} = (1,1,1,0,1)$	7	$v_6, v_9, v_{11}, v_{13}, v_{15}, v_{16}, v_{18}$
$v_{18} = (1,1,1,1,0)$	7	$v_7, v_{10}, v_{11}, v_{14}, \ldots, v_{17}$

TABLE 7. five-dimensional 0/1-polytope with 18 vertices and 75 edges. Vertex v_{11} achieves the maximal degree 17.

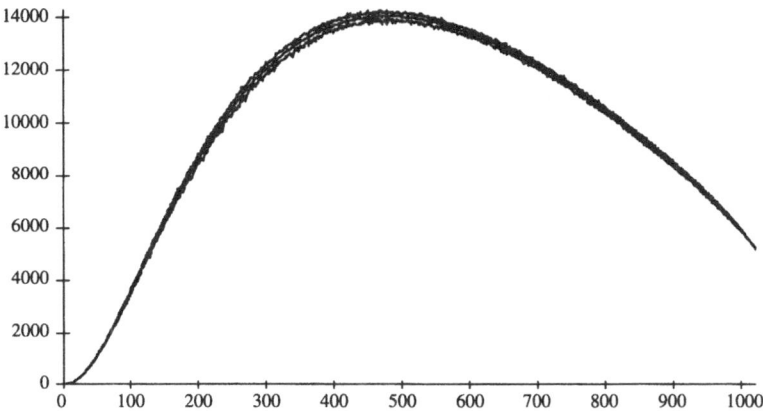

FIGURE 4. Average number of edges of 10-dimensional random 0/1-polytopes over the number of vertices.

or facets (moreover the 5-cube is the only simple five-dimensional 0/1-polytope with 10 facets). Most recently Kaibel and Wolff showed that in fact all simple 0/1-polytopes are the (combinatorial) product of simplices, see [16].

$$
\begin{array}{ll}
v_1 = (0,0,0,0,0) & v_6 = (1,0,1,1,1) \\
v_2 = (0,0,0,0,1) & v_7 = (1,1,0,1,1) \\
v_3 = (0,0,0,1,0) & v_8 = (1,1,1,0,1) \\
v_4 = (0,0,1,0,0) & v_9 = (1,1,1,1,0) \\
v_5 = (0,1,0,0,0) & v_{10} = (1,1,1,1,1)
\end{array}
$$

TABLE 8. Vertices of a simplicial five-dimensional 0/1-polytope with the maximal number of 10 vertices, 40 edges and 32 facets.

For simplicial 0/1-polytopes the situation is more interesting, although there exists again an extremal 0/1-polytope maximizing at the same time the number of vertices (10), edges (40) and facets (32), respectively. Table 8 gives the vertex set of one such 0/1-polytope. As already mentioned above, all in all we found 7 equivalence classes of this type, with 3008 0/1-polytopes in total.

If we in addition require a simplicial 0/1-polytope to be 2-neighbourly (see Section 4.1) then we get as extremal values 8 vertices, 28 edges and 20 facets.

Let us close this section with a remark: All the "special" 0/1-polytopes of this paper (Tables 4,6,7,8 and 9) are available in the POLYMAKE library (cf. [10]).

4. Higher dimensions

In this section we give some results for 0/1-polytopes of dimension $d \geq 6$, obtained as spin-off from the complete enumeration of dimension 5.

4.1. 2-neighbourly 0/1-polytopes

As mentioned in the last section a 2-neighbourly 0/1-polytope is a polytope where any pair of vertices is adjacent. This means that the convex hull of a 2-neighbourly 0/1-polytope with k vertices has $\frac{k(k-1)}{2}$ edges, or, in other words, it represents a complete graph. Clearly 2-neighbourly 0/1-polytopes are worst case examples for algorithms based on the edge skeleton of their convex hull. Thus for fixed dimension d one may ask for the maximal number $N_2(d)$ of vertices of the d-cube for which a 2-neighbourly 0/1-polytope exists.

Obviously the d-simplex is a 2-neighbourly polytope, i.e. $N_2(d) \geq d+1$. On the other hand "one half" of a 2-neighbourly 0/1-polytope for the d-cube yields a 2-neighbourly 0/1-polytope for the $(d-1)$-cube. And adding one vertex to a 2-neighbourly 0/1-polytope for the $(d-1)$-cube gives a 2-neighbourly 0/1-polytope for the d-cube. Thus $N_2(d-1)+1 \leq N_2(d) \leq 2N_2(d-1)$.

We have $N_2(d) = d+1$ for $d = 1\ldots 3$, and it is well known that $N_2(4) = 6$, i.e. there exists a 2-neighbourly 0/1-polytope for dimension 4 that is not a simplex, a counter-intuitive phenomenon which has no analogue for $d \leq 3$.

Complete enumeration gives the maximal value for $d = 5$ and $d = 6$. To this end we utilize a result of the last section: $N_2(5) = 10$, where only one equivalence

$d = 4$	$d = 5$	$d = 6$
$v_1 = (0,0,0,0)$	$v_1 = (0,0,0,0,0)$	$v_1 = (0,0,0,0,0,0)$
$v_2 = (0,0,0,1)$	$v_2 = (0,0,0,0,1)$	$v_2 = (0,0,0,0,0,1)$
$v_3 = (0,0,1,0)$	$v_3 = (0,0,1,1,0)$	$v_3 = (0,0,0,1,1,0)$
$v_4 = (0,1,0,0)$	$v_4 = (0,1,0,1,1)$	$v_4 = (0,0,1,0,1,1)$
$v_5 = (1,0,0,1)$	$v_5 = (0,1,1,1,0)$	$v_5 = (0,0,1,1,1,0)$
$v_6 = (1,1,1,0)$	$v_6 = (1,0,1,0,1)$	$v_6 = (0,1,0,1,0,1)$
	$v_7 = (1,0,1,1,1)$	$v_7 = (0,1,0,1,1,1)$
	$v_8 = (1,1,0,0,0)$	$v_8 = (0,1,1,0,0,0)$
	$v_9 = (1,1,0,1,1)$	$v_9 = (0,1,1,0,1,1)$
	$v_{10} = (1,1,1,0,0)$	$v_{10} = (0,1,1,1,0,0)$
		$v_{11} = (1,0,0,0,0,0)$
		$v_{12} = (1,0,1,1,0,0)$
		$v_{13} = (1,1,0,0,1,0)$

TABLE 9. Vertices of maximal 2-neighbourly 0/1-polytopes for dimensions 4 to 6.

dimension	1	2	3	4	5	6	7	8	9	10
best example	2	3	4	6	10	13	≥ 18	≥ 25	≥ 33	≥ 44
upper bound	2	3	4	6	10	13	24	48	96	192

TABLE 10. 2-neighbourly 0/1-polytopes with maximal number of vertices for dimensions 1 to 10.

class provides a 2-neighbourly 0/1-polytope with 10 vertices. For dimension $d = 6$ random vertex sets give a lower bound of $N_2(6) \geq 13$. From the above discussion it follows that any vertex set of C^6 with $N_2(6) \geq 14$ must contain a subset of C^5 with 7 or more vertices. We thus considered all 447 different 2-neighbourly vertex sets of C^5 with at least 7 vertices and completed them with vertices on the "opposite half" of C^6. A complete enumeration of all resulting 6-dimensional 2-neighbourly 0/1-polytopes (about half a million) showed $N_2(6) = 13$.

Table 9 gives the vertex sets for the maximal 2-neighbourly 0/1-polytopes for dimensions 4 to 6. It is interesting to note that the 0/1-polytope for dimension 5 is 'one half' of the 6-dimensional example, but does itself only contain subsets of C^4 with at most 5 vertices. In fact no five-dimensional 2-neighbourly 0/1-polytope that has a subset of 6 vertices from C^4 has more than 9 vertices.

For higher dimensions we made incomplete enumerations. E.g. for dimension $d = 7$ we used a similar method as for dimension 6, getting the lower bound $N_2(7) \geq 18$. The investigations also showed that if there exists a vertex set of C^7 with $N_2(7) \geq 19$ then no 6-dimensional subcube contains more than 12 of its vertices. This yields an upper bound of $N_2(7) \leq 24$. Table 10 gives $N_2(d)$ for dimensions $d = 1 \ldots 10$. For general dimension there is an exponential lower bound for $N_2(d)$, see [18].

4.2. Edge skeleton of 0/1-polytopes

There are several practical applications, such as clustering binary data or $\{0,1\}$-string searching (see e.g. Chapters 7 and 8 in [1]), which are based on the edge skeleton of high-dimensional 0/1-polytopes. Typical input parameters for these tasks might be $d = 100 \ldots 200$ with $n = 1000 \ldots 50000$ vertices. Thus the number of vertices is rather small compared to the number of "corners" of the d-cube. To get some insight of the structure of high-dimensional edge skeletons with few vertices we investigated random vertex sets for dimension $d = 15$ of up to 240 vertices.

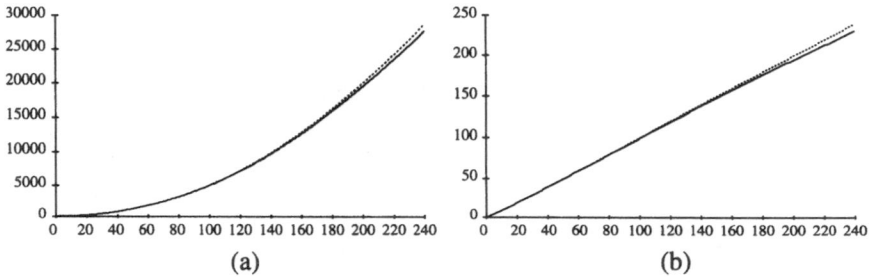

(a)   (b)

FIGURE 5. Edge skeleton of 0/1-polytopes for dimension $d = 15$ and $n = 1 \ldots 240$ vertices: (a) average number of edges (solid) and upper bound $n(n-1)/2$ (dotted) (b) average edge degree (solid) and upper bound $n-1$ (dotted).

For dimension $d = 10$ Figure 4 already shows that for a small number of vertices the maximal number of edges of the convex hull increases quadratically. Figure 5(a) shows the average size of the edge skeleton for dimension $d = 15$ and $n = 1 \ldots 240$ vertices, compared to the upper bound of $n(n-1)/2$. It can be seen that the size of the edge skeleton almost matches the upper bound.

Figure 5(b) gives the average edge degree and the upper bound $n-1$. Again the curves are very similar, i.e. a random 0/1-polytope with the given parameters is very likely to be a good approximation of a 2-neighbourly 0/1-polytope with the same number of vertices. This indicates that algorithms based on the edge skeleton of the convex hull of high-dimensional 0/1-polytopes are most likely to attain their worst case behavior.

Figure 6 gives the maximal average edge degree and the maximal number of edges of d-dimensional 0/1-polytopes for dimension $d = 1 \ldots 10$. To this end we considered random samples of 100 0/1-polytopes for each dimension and each number of vertices. In correspondence with Figure 5 these curves show that the number of edges and the average edge degree grows significantly with increasing dimensions.

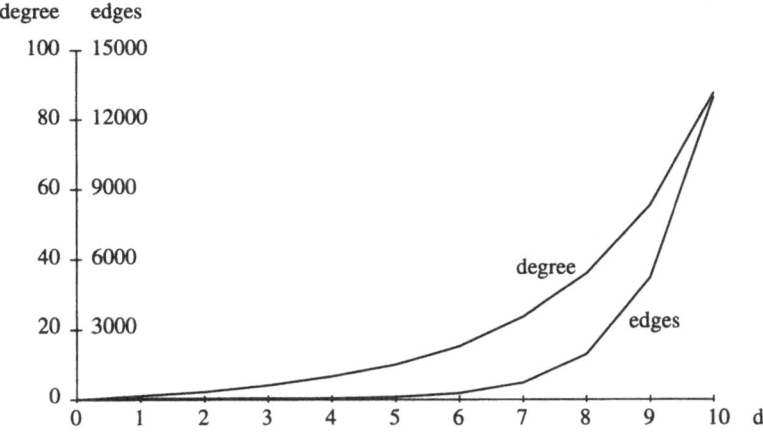

FIGURE 6. Maximal average edge degree and maximal number of edges of 0/1-polytopes for dimension $d = 1 \ldots 10$.

Acknowledgements. We would like to thank Franz Aurenhammer, Günter Rote and Günter M. Ziegler for several discussions and helpful comments on the presented topic.

References

[1] O. AICHHOLZER, *Combinatorial & Computational Properties of the Hypercube*, Ph.D. Thesis, TU-Graz, Austria, 1997

[2] O. AICHHOLZER, F. AURENHAMMER, *Classifying Hyperplanes in Hypercubes*, SIAM J. Discrete Math, Vol. 9, No.2, pp 225–232, 1996

[3] D. AVIS, D. BREMNER, R. SEIDEL, *How Good are Convex Hull Algorithms?*, Computational Geometry: Theory and Applications, Vol. 7, pp 265–302, 1997

[4] E.R. BERLEKAMP, *Algebraic Coding Theory*, McGraw Hill, New York, 1968

[5] N.G. DE BRUIJN, *Pólya's Theory of Counting*, in E. F. Beckenbach (ed.), Applied Combinatorial Mathematics, Wiley, NY 1964

[6] B. BÜELER, A. ENGE, K. FUKUDA, *Exact Volume Computation for Polytopes – A Practical Study*, in this volume, pp. 131–154

[7] W. BURNSIDE, *Theory of Groups of Finite Order*, Cambridge University Press, 1911

[8] H.S.M. COXETER, *Regular Polytopes*, Dover Publications, New York, 1963/73

[9] T. FLEINER, V. KAIBEL, G. ROTE, *Upper bounds on the maximal number of facets of 0/1-polytopes*, European J. Combinatorics, Vol. 21, pp 121–130, 2000

[10] E. GAWRILOW, M. JOSWIG, *Polymake: a framework for analyzing convex polytopes*, in this volume, pp. 43–74

[11] E. GAWRILOW, M. JOSWIG, *POLYMAKE: a software package for analyzing convex polytopes*, http://www.math.tu-berlin.de/diskregeom/polymake/doc/

[12] E.N. GILBERT, *Gray Codes and Paths on the n-Cube*, Bell Systems Tech. J. 37, pp 1–12, 1958

[13] B. GRÜNBAUM, *Convex Polytopes*, Interscience, New York, 1967

[14] M. HENK, J. RICHTER-GEBERT, G.M. ZIEGLER, *Basic Properties of Convex Polytopes*, CRC Handbook on Discrete and Computational Geometry (J.E. Goodman, J.O.'Rourke, eds), pp 243–270, 1997

[15] K. JACOBS, *Einführung in die Kombinatorik*, de Gruyter, Berlin New York, 1983

[16] V. KAIBEL, M. WOLFF, *Simple 0/1-Polytopes*, European J. Combinatorics, Vol. 21, pp 139–144, 2000

[17] U.H. KORTENKAMP, *Small 0/1-Polytopes with Many Facets*, http://www.math.tu-berlin.de/~hund/01-Olympics.html

[18] U.H. KORTENKAMP, J. RICHTER-GEBERT, A. SARANGARAJAN, G. M. ZIEGLER, *Extremal Properties of 0/1-Polytopes*, Discrete & Computational Geometry Vol. 17, pp 439–448, 1997

[19] J.H. VAN LINT, R. M. WILSON, *A Course in Combinatorics*, Cambridge University Press, 1992

[20] G. PÓLYA, *Kombinatorische Anzahlbestimmungen für Gruppen, Graphen und chemische Verbindungen*, Acta Mathematica Vol. 68, pp 145–254, 1937

[21] N.J.A. SLOANE, *A Short Course on Error Correcting Codes*, CISM Courses and Lectures 188, Springer, Wien, New York, 1975

[22] G.M. ZIEGLER, *Lectures on Polytopes*, Graduate Texts in Mathematics 152, Springer-Verlag, 1995; Revised printing 1998

[23] G.M. ZIEGLER, *Lectures on 0/1-Polytopes*, in this volume, pp. 1–41

[24] G.M. ZIEGLER, *Recent Progress on Polytopes*, in: "Advances in Discrete and Computational Geometry" (B. Chazelle, J.E. Goodman, R. Pollack, eds.), Contemporary Mathematics, Amer. Math. Soc., Providence, 1998

[25] G.M. ZIEGLER, *Polytopes and Optimization: Recent Progress and Some Challenges*, GMÖOR Newsletter 3/4, pp 3–12, 1996

Oswin Aichholzer
Institute for Theoretical Computer Science
Graz University of Technology
Klosterwiesgasse 32/2
A-8010 Graz, Austria
oaich@igi.tu-graz.ac.at

Exact Volume Computation for Polytopes: A Practical Study

Benno Büeler, Andreas Enge and Komei Fukuda

Abstract. We study several known volume computation algorithms for convex d-polytopes by classifying them into two classes, triangulation methods and signed-decomposition methods. By incorporating the detection of simplicial faces and a storing/reusing scheme for face volumes we propose practical and theoretical improvements for two of the algorithms. Finally we present a hybrid method combining advantages from the two algorithmic classes. The behaviour of the algorithms is theoretically analysed for hypercubes and practically tested on a wide range of polytopes, where the new hybrid method proves to be superior.

1. Introduction

A *convex polytope* P is the convex hull $\mathrm{conv}(V)$ of a finite set $V = \{v_1, v_2, \ldots, v_n\}$ of points in $\mathbb{R}^d$. Equivalently, it is a bounded subset of $\mathbb{R}^d$ which is the intersection of a finite set of half spaces. We shall omit 'convex' since we only deal with such polytopes, and use *d-polytope* to mean a d-dimensional polytope. When $P = \mathrm{conv}(V)$, V is called a *vertex representation* or simply $\mathcal{V}$-*representation* of P. When $P = \{x | A\, x \le b\}$ for some $m \times d$ real matrix A and m-vector b, the pair (A, b) is called a *halfspace representation* or simply $\mathcal{H}$-*representation* of P. In this paper we treat the *volume computation problem* as to compute the volume $\mathrm{Vol}(P)$ of a polytope P given by both $\mathcal{V}$- and $\mathcal{H}$-representations. It seems that the theoretical complexity of this problem is unknown and not easy to evaluate, although the same problem when only one (either $\mathcal{V}$- or $\mathcal{H}$-) representation is provided is known to be #P-hard, see [5]. Main reasons for considering this problem setting are (1) the volume computation appears to be already hard with two representations given, (2) the transformation between the two representations is itself a hard fundamental problem which has been studied both theoretically and practically [6, 10, 1, 3, 9], and (3) many different algorithms, some requiring only one and some requiring both representations, can be compared on the same ground in our setting.

This research was partially supported by the ETHZ-EPFL joint research project on optimization and geometric computation, Switzerland. See
http://www.ifor.math.ethz.ch/ifor/staff/fukuda/OGC_home/OGC.html.

The relevance of volume computation is nicely demonstrated in the excellent survey [11], where also several basic approaches for volume computation are discussed in depth. In this paper we consider five methods and classify them into two groups. Triangulation methods on the one hand decompose the polytope into simplices for which the volume is easily computed and summed up; members of this group are the boundary triangulation, Delaunay triangulation and Cohen & Hickey's triangulation [4]. Signed decomposition methods, on the other hand, decompose a given polytope into signed simplices such that the signed sum of their volumes is the volume of the polytope; in this group we consider Lasserre's [13] and Lawrence's [15] methods. As explained in Section 2, these methods are in a sense dual to triangulation methods. A basic algorithm excluded from our investigations is the incremental construction of a triangulation using the beneath-beyond method [16, 7] for convex hull computation (see also [11, Section 4]). Concerning random approximation algorithms (see [12]) we know so far of no implementation and have therefore ignored this highly interesting new approach.

During an earlier stage of our study, we observed that no method described in the literature works efficiently on a wide range of polytopes. In fact, each of the investigated methods is hopelessly impractical for either the class of simple polytopes or the class of simplicial polytopes. For example, all triangulation methods work poorly for simple polytopes, and Lawrence's signed decomposition method crawls for simplicial polytopes. This is not surprising, since the sizes of a triangulation and a signed decomposition of the same polytope can be drastically different, see Section 2.4. This observation was our major motivation to revise some of the known methods so that they behave more uniformly and efficiently for a broader range of polytopes. In particular, in Lasserre's method we found a significant potential for improvements. These include storing and reusing intermediate results (in this case volumes of some polytope faces) and the efficient detection of empty faces. Cohen & Hickey's triangulation profits considerably from some more obvious improvements. Our modifications shall be related in detail in Section 4. Their success incited us to design a hybrid algorithm, combining the strong sides of both Cohen & Hickey's and Lasserre's methods. This new algorithm has an overall convincing performance, to which the laurels can be awarded in our competition.

While experiments showed differences among the algorithms in CPU-time and memory up to several orders of magnitude, the theoretical analysis is difficult. One major source is related to the difficulty of classifying polytopes, for example by a satisfactory measure for 'how simple' or 'how simplicial' a polytope is. Another difficulty is introduced by the recursive or iterative structure of the algorithms; it is usually not only the structure of the original polytope, but of all intermediate polytopes generated in the solution process which determine the complexity. Despite these difficulties we give complexity estimates for some concrete algorithms and polytopes accepting their limited relevance and hoping that it is a first step in the explanation of the real algorithmic behaviour for a wider class of polytopes. In this state experiments may prove useful to gain a better understanding of poly-

topal structures and the behaviour of different algorithmic concepts. For a broader
survey on complexity results see e.g. [11].

It is important to note that the algorithms differ in the input they require;
because the transformation $V \rightleftarrows H$ can be prohibitively costly for some polytopes,
this can influence decisively the choice of algorithms. Though highly depending
on the concrete structure of the polytope under consideration, one can expect
tractability on workstations for polytopes with up to 10^2 hyperplanes and 10^4
vertices, or, 10^4 hyperplanes and 10^2 vertices, in dimension up to 15, requiring
memory up to the range of 10^2–10^3 MB, see Table 3.

The paper is organised as follows. In Section 2 we discuss two basic ideas
of volume computation together with their dual relation, allowing a deeper understanding of the complexity observed. Section 3 is devoted to the five basic
algorithms we have chosen from the literature. Then, in Section 4, we propose a
number of improvements for Cohen & Hickey's and Lasserre's methods and present
a hybrid algorithm. We conclude in Section 5 by presenting experimental results for
seven classes of polytopes. As a by-product of our work, the code of all algorithms
discussed is publicly available; the sources are given in the Appendix.

For a theoretical analysis, we assume a computer device with arbitrary precision. Namely we assume that all elementary arithmetic operations require an
amount of time which is independent of the number of digits needed in the representation of the numbers. In such a computational model, the time complexity
is proportional to the number of elementary arithmetic operations. While this assumption makes complexity analysis much simpler, it has an obvious drawback
when some algorithms require much higher precision than others. We shall discuss
the numerical issue independently for each algorithm in the Sections 3 and 5.

2. Basic algorithmic concepts and duality

All known algorithms for exact volume computation decompose a given polytope
into simplices, and thus they all rely, explicitly or implicitly, on the volume formula
of a simplex:

$$\mathrm{Vol}(\Delta(v_0, \ldots, v_d)) = \frac{|\det(v_1 - v_0, \ldots, v_d - v_0)|}{d!},$$

where $\Delta(v_0, \ldots, v_d)$ denotes the simplex in $\mathbb{R}^d$ with vertices $v_0, \ldots, v_d \in \mathbb{R}^d$. There
are two types of methods for volume computation, depending on how a given
polytope P is decomposed into simplices. As we shall explain later, these two
algorithm classes are related by polarity of convex polytopes.

2.1. Triangulation methods

A *triangulation* of a d-polytope P is a set $\{\Delta_i : i = 1, \ldots, s\}$ of d-simplices such
that $P = \cup_{i=1}^s \Delta_i$, and no distinct simplices have an interior point in common.

Then the volume of P is simply the sum of the volumes of the simplices:

$$\text{Vol}(P) = \sum_{i=1}^{s} \text{Vol}(\Delta_i). \tag{1}$$

Figure 1 illustrates a triangulation[1] using an interior point e; it is often used when the boundary of a polytope can be easily triangulated or is already triangulated as in the case of simplicial polytopes. Besides such a boundary triangulation method we will also consider the Delaunay triangulation and Cohen & Hickey's combinatorial triangulation by dimensional recursion. We postpone a detailed description of these methods to Section 3.1. A first important difference is that the former two methods need only a $\mathcal{V}$-representation while the last method requires both the $\mathcal{V}$- and $\mathcal{H}$-representations.

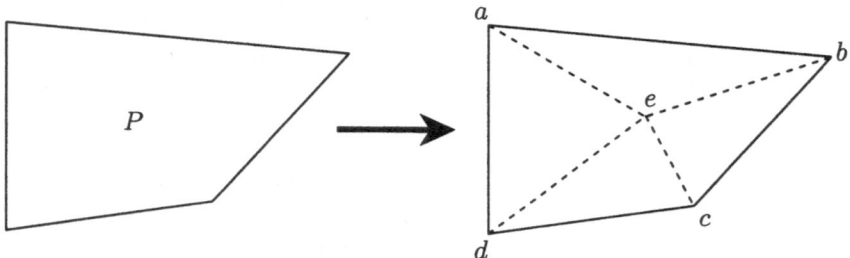

FIGURE 1. Boundary triangulation

2.2. Signed decomposition methods

Instead of triangulating a polytope P, one can decompose P into 'signed' simplices whose signed union is exactly P. More specifically, we represent P as a signed union of simplices Δ_i, $i = 1, \ldots, s$,

$$P = \biguplus_{i=1}^{s} \sigma_i \Delta_i, \tag{2}$$

where σ_i is either $+1$ or -1 and the right hand side is considered as a multi-set whose elements have (possibly negative) multiplicity. Equation (2) means that each point x in $\mathbb{R}^d$ not on the boundary of any of the Δ_i's has the following property: If x is in P, it appears in exactly one more positive Δ_i than in negative Δ_i's, and otherwise it appears equally often in positive and in negative Δ_i's. Then the volume of P is

$$\text{Vol}(P) = \sum_{i=1}^{s} \sigma_i \, \text{Vol}(\Delta_i). \tag{3}$$

---

[1]This definition is slightly weaker than the usual one that requires the intersection of any two simplices being their common face. This additional property is satisfied by most known constructions of triangulations, but not necessary for the purpose of volume computation.

In this paper we consider two signed decomposition methods, Lawrence's decomposition as illustrated in Figure 2, and Lasserre's decomposition. These methods are presented in detail in Section 3.2. Fundamental to Lawrence's method for sim-

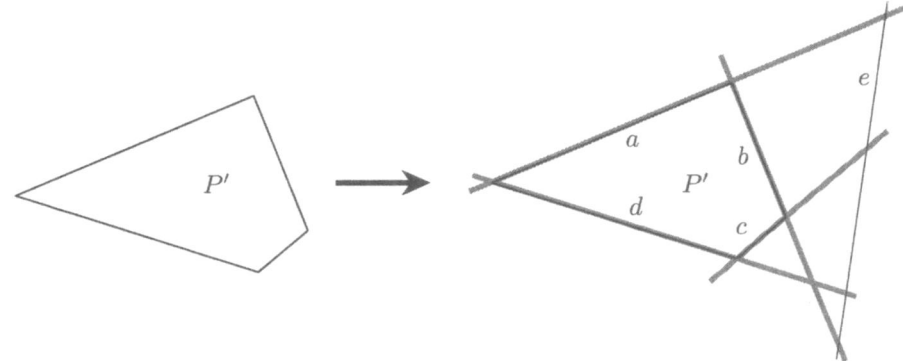

FIGURE 2. Lawrence's decomposition scheme

ple polytopes is the introduction of an extra hyperplane e which is not parallel to any edge of the polytope. The simplices in the signed decomposition are generated by passing once through each vertex, and taking the simplex built by the vertex-touching hyperplanes together with e. In the example shown in Figure 2 the decomposition contains two positive simplices, ade and cbe, and two negative simplices, abe and cde, where xyz denotes the unique simplex determined by the hyperplanes x, y and z.

Note that Lasserre's method requires only the $\mathcal{H}$-representation, while Lawrence's method requires both the $\mathcal{V}$- and the $\mathcal{H}$-representations.

2.3. Duality

Assume that the origin lies in the interior of a polytope P; then the polar polytope P' is defined by $P' := \{x \mid y^T x \leq 1 \ \forall y \in P\}$. Note that the vertices are mapped onto the facets and vice versa as is illustrated in Figure 3. Filliman [8] showed that there is a dual relation between a triangulation of a polytope and a signed decomposition of its polar. While the left polytope is triangulated into the four simplices $\Delta(abe)$, $\Delta(bce)$, $\Delta(cde)$ and $\Delta(ade)$, the polar P' is decomposed into signed simplices determined by the same triples. The sign of each simplex xyz in the decomposition of P' is determined by the *separation parity* of the corresponding simplex $\Delta(xyz)$ in the triangulation of P, which is the parity of the number of facet-defining hyperplanes of $\Delta(xyz)$ which separate the simplex from the origin. The simplex $\Delta(bce)$ has even parity ($+1$) since it has two facet-defining lines be and ce separating it from the origin. The simplex cde has odd parity (-1) since it has only one such line de. When the parity is odd, the corresponding simplex in the signed decomposition of the polar is negative. Thus we have two positive simplices cbe, ade and two negative simplices cde, abe.

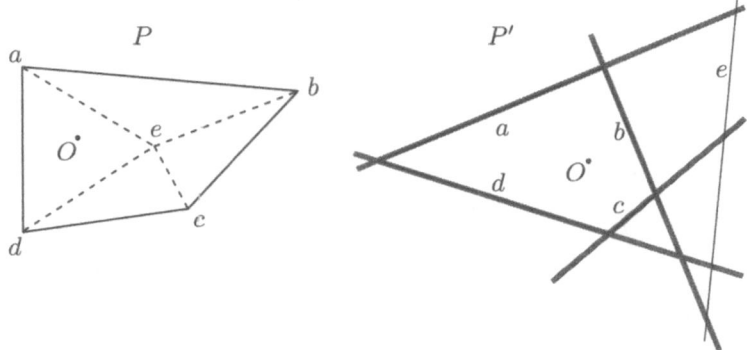

FIGURE 3. Filliman's duality

More generally Filliman showed:

> Let P be a d-polytope containing the origin in its interior and let $\mathcal{T}$ be a triangulation of P such that the origin is not contained in the union of hyperplanes spanned by vertices of the triangulation. Then the triangulation induces a signed decomposition of the polar P' of P in such a way that each simplex $\Delta(a_0, a_1, \ldots, a_d)$ is in $\mathcal{T}$ if and only if the unique bounded simplex determined by the corresponding hyperplanes in the polar appears as a simplex in the decomposition. The sign of each simplex in the signed decomposition is determined by its separation parity in the triangulation.

Observe that the assumption on the location of the origin is important in defining the separation parity properly.

2.4. Duality and efficient volume computation

What does duality tells us for volume computation? Should one triangulate or determine a signed decomposition of a polytope to compute the volume? While these questions do not have a simple answer, they raise a basic point that should be kept in mind when we design or use volume computation algorithms. That is, for any fixed polytope, the size of a triangulation and that of a signed decomposition might be drastically different.

To illustrate this claim the unit d-hypercube C^d of volume 1 and its polar, the cross polytope, can serve as an example. C^d has 2^d vertices and $2d$ facets, and consequently Lawrence's signed decomposition produces 2^d simplices. On the other hand, a simple recursive triangulation of C^d (see Cohen & Hickey in Section 3 for details) has exactly $d!$ simplices. Are there any triangulations with much fewer simplices? To estimate a simple lower bound for the number of simplices in a triangulation we can use Hadamard's upper bound for the maximum volume of simplices inscribed in the unit hypercube, $d^{d/2}/d!$. Although the resulting lower bound for the number of simplices, $O((d/2)!)$, might be far from the unknown minimum number, the growth in the number of simplices is much faster for any triangulation method compared to Lawrence's signed decomposition. If we take the cross polytope, the situation is completely reversed.

When we know the number of simplices generated by an algorithm, this yields a trivial lower bound for the complexity of the algorithm. The above observation shows that such a lower bound can be prohibitively high, while there might be much better alternatives.

Indeed, in the case of the methods drawn from the literature we experimentally found a behaviour following exactly these hypotheses: All triangulation methods fail for hypercubes and convince for cross polytopes, while Lawrence's signed decomposition method behaves in the opposite way. This was a strong motivation to develop a hybrid algorithm that combines the advantages of different methods.

It should be noted that Lasserre's method to be described in the next section is a signed decomposition, but its complexity does not relate well with those of the other algorithms. The only case we could analyse easily was the hypercube, for which the complexity is $O(d!)$, whence the algorithm behaves more like triangulation methods, see Section 3.2.

3. Basic algorithms

In this section representatives of triangulation methods and signed decomposition methods described in the literature are presented. To allow a theoretical comparison we give some complexity analyses. While it is perfectly possible to give general upper bounds as in Section 2.4, these bounds are usually far from being tight. Hence, they hardly explain why an algorithm behaves differently on various polytope classes. On the other hand, it seems to be very difficult to obtain tight complexity results for a specific class of polytopes, since the volume computation algorithms depend on associated geometric and combinatorial structures of a given d-polytope that are not fully understood. So we decided to analyse the performance of the different methods on hypercubes, whose geometric and combinatorial structures are quite simple. We expect (and this could be observed in experiments, see Section 5) that the behaviour of the different methods on hypercubes is typical of a broader class of polytopes, namely the simple or near-simple polytopes, so that by duality, our analysis should well enlighten the practical performance of the algorithms.

3.1. Triangulation methods

DELAUNAY TRIANGULATION.   The geometric idea behind a Delaunay triangulation of a d-polytope is to 'lift' it on a paraboloid in dimension $d+1$. This is done on the level of vertices $v \to \bar{v}$, $(x_1, \dots, x_d) \mapsto (x_1, \dots, x_d, \sum_{i=1}^{d} x_i^2)$. If the resulting vertices are in general position, the convex hull of $\bar{V}$ yields a triangulation, whose 'lower half' is a Delaunay triangulation. In our experiments the underlying convex hull algorithm uses the 'beneath-beyond' method (see [11]). This method requires only the $\mathcal{V}$-representation.

BOUNDARY TRIANGULATION. Even if P itself is not simplicial, lexicographic perturbation (either symbolically or numerically) of the vertices leads to a simplicial polytope. Computing the convex hull of the perturbed points and interpreting the result in terms of the original vertices leads to a triangulation of the boundary which, by linking with a fixed interior point, yields a triangulation of P. For the convex hull computation we chose the reverse search algorithm [1], where only the $\mathcal{V}$-representation is required as input.

TRIANGULATION BY COHEN & HICKEY. This recursive scheme triangulates a d-polytope P by choosing any vertex $v \in P$ as apex and connecting it with the $d - 1$-dimensional simplices resulting from a triangulation of all facets of P not containing v, see [4]. In the following, the notation e^k stands for some k-dimensional face of P. Let η be a map which associates to each face one of its vertices. Then the pyramids with apex $\eta(e^d)$ and bases among all facets e^{d-1} with $\eta(e^d) \notin e^{d-1}$ form a dissection of the polytope. Applying this scheme recursively to all e^{d-1} under consideration results in a set of decreasing chains of faces, $\mathcal{C} := \{(e^d, \ldots, e^1, e^0) \mid e^k \supset e^{k-1} \text{ and } \eta(e^k) \notin e^{k-1} \text{ for all } 1 \leq k \leq d\}$. Then the set of corresponding simplices $\{\Delta(\eta(e^0), \ldots, \eta(e^d)) \mid (e^d, \ldots, e^0) \in \mathcal{C}\}$ is a triangulation of P. To implement this recursive method extensive use of the double description as $\mathcal{V}$- and $\mathcal{H}$-polytope is made by representing all faces as sets of vertices. Based on this combinatorial representation, we compute the list of possible e^{k-1} from a given e^k by intersecting the set of vertices of e^k with the facets e^{d-1} of P not containing the vertex $\eta(e^k)$. We do this by a binary look-up of the vertices of e^k in the set of vertices of e^{d-1}. To prevent the multiple generation of the same face in degenerate cases, a list is maintained containing all faces of e^k generated so far; only faces which are not in this list are accepted. In the sequel we call this basic algorithmic scheme $C\&H$.

Note that in the case of C&H compared to a boundary triangulation all simplices in the facets containing the apex v are eliminated and thereby the number of simplices is usually reduced.

As an example consider Figure 4 where η assigns to each face its vertex with the lowest number, so $\eta(P) = 0$. Now all facets which do not contain the vertex 0 are examined, that is, II, III and IV. The scheme is now applied to facet II with $\eta(\text{II}) = 1$. II is intersected with all facets not containing the vertex 1, these are III, IV and V. The intersections with IV and V are empty, so these recursion branches are unsuccessful. The intersection with III yields the vertex 2, and the fixed vertices 0, 1 and 2 form a first simplex. The other simplices obtained from III and IV are also marked in the figure.

Proposition 1. *The time complexity of C&H applied to a d-hypercube is $O(d^3 d!)$.*

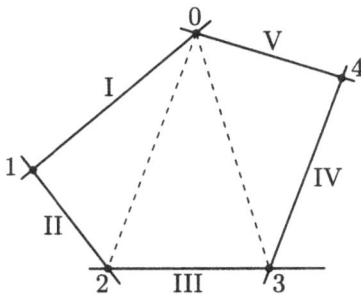

FIGURE 4. Cohen & Hickey's triangulation scheme

Proof. There are $2^{d-k}\binom{d}{k}$ faces of dimension k, each of which contains 2^k vertices. The recursion tree resembles the face lattice where some faces do not appear whereas others are present multiple times. The (unique) root is the original d-polytope P, and in each recursion level k *some* of the incident $(k-1)$-dimensional faces e^{k-1} are included. Denote by s_k the number of faces of dimension k appearing in the recursion tree; from the construction we know $s_d = 1$. In recursion k we intersect a face e^k with all facets not containing $\eta(e^k)$, and since there are k such facets we have $s_{k-1} = ks_k$, implying $s_k = \frac{d!}{k!}$. Hence the number of simplices, s_0, equals $d!$. Because for each simplex a determinant has to be computed, the overall complexity bound derived so far is $O(d^3 d!)$, assuming conventional matrix computation techniques. In the rest of the proof we verify that the additional computational burden of the algorithm per simplex is bounded by $O(d^3)$. Suppose that the intersection of e^k with a facet e^{d-1} is done by a binary look-up of the vertices of e^k in e^{d-1}, resulting in $O(|e^k|\log|e^{d-1}|) = O((d-1)2^k) \subseteq O(d2^k)$ comparisons, where the notation $|e^k|$ designates the number of vertices in the face e^k. To check further whether the newly determined face e^{k-1} appears in the list of at most $k-1$ faces already derived from e^k requires another $O((k-1)|e^{k-1}|) \subseteq O(d2^k)$ comparisons. The facet-intersection and the maintenance of the list have to be done for each facet, resulting in a total complexity of $O(\sum_{k=1}^{d} s_k k(d2^k + d2^k))$, where

$$\sum_{k=1}^{d} s_k k d 2^k = dd! \sum_{k=1}^{d} \frac{2^k}{k!} k < 2dd! \sum_{k=0}^{\infty} \frac{2^k}{k!} k = 2e^2 dd!.$$

Thus the additional computational burden per simplex is bounded by $O(d)$ which can be neglected compared to the determinant computation. □

3.2. Signed decomposition methods

LASSERRE'S RECURSIVE ALGORITHM.   The volume is computed via a recursive scheme based on Euler's formula for homogeneous functions. Denote by $\mathrm{Vol}(d, A, b)$ the volume sought, by a_i the ith row of A, and by $\mathrm{Vol}_i(d-1, A, b)$ the volume in the appropriate $\mathbb{R}^{d-1}$ subspace defined by the face $P \cap \{a_i^T x = b_i\}$. Then the volume of P can be expressed by the following recursive structure:

$$\mathrm{Vol}(d, A, b) = \frac{1}{d} \sum_{i=1}^{m} \frac{b_i}{\|a_i\|} \mathrm{Vol}_i(d-1, A, b).$$ (4)

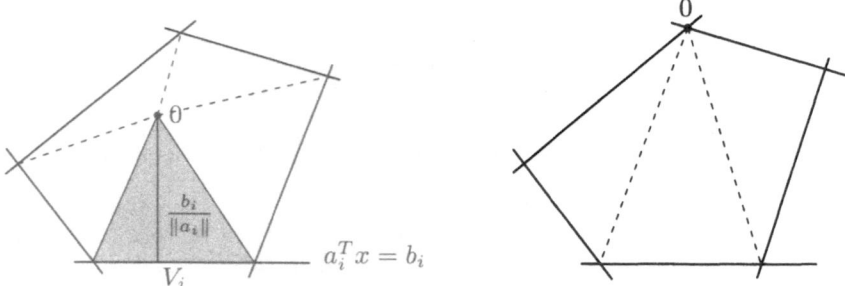

FIGURE 5. Triangulation induced by Lasserre's method if the origin is inside a facet (left) or if it coincides with a vertex (right). The volume $\frac{b_i}{d\|a_i\|} \mathrm{Vol}_i(d-1, A, b)$ of a single triangle is gray shaded in the left graph.

Now (4) is not yet an implementable recursion scheme because the terms in the sum have to be evaluated in appropriate lower dimensional spaces. To compute $\mathrm{Vol}_i(d-1, A, b)$ Lasserre [13] suggests a projection scheme as follows. For the constraint to be fixed choose a pivot element $a_{ij} \neq 0$ and set $x_j = (b_i - \sum_{k \neq j} a_{ik} x_{ik})/a_{ij}$. Substituting x_j in this way defines a new system $\tilde{A}\tilde{x} \leq \tilde{b}$ with $m-1$ constraints and $\tilde{d} := d-1$ variables for which we have $\mathrm{Vol}_i(d-1, A, b) = (\|a_i\|/a_{ij}) \mathrm{Vol}(\tilde{d}, \tilde{A}, \tilde{b})$. Geometrically $\mathrm{Vol}(\tilde{d}, \tilde{A}, \tilde{b})$ is the volume of $P \cap \{a_i^T x = b_i\}$ projected onto the subspace $\{x_j = 0\}$. Finally, using (4) gives the recursive scheme we implemented:

$$\mathrm{Vol}(d, A, b) = \frac{1}{d} \sum_{i=1}^{m} \frac{b_i}{a_{ij}} \mathrm{Vol}(\tilde{d}, \tilde{A}, \tilde{b}).$$ (5)

At the bottom of the recursion tree, the volume of a line segment bounded by (at most) $m - d + 1$ constraints has to be computed. It is of critical importance in this formula that no constraint appears more than once in any (recursively generated) face. Even though this is usually given for the input, it turns out that in many interesting examples multiple constraints occur in lower dimensional faces. In the process of eliminating identical constraints non-identical parallel constraints are treated simultaneously, where two cases are possible: either the intersection of two parallel halfspaces is non-empty or empty. While we can stop the recursion branch in the latter case, we check in the former case whether the normal vectors point in the same direction and if so drop the 'outer' of the two parallel hyperplanes.

For hypercubes this implies that in dimension k there are $2k$ constraints in the reduced system.

Looking at (5) another simple improvement is possible by making as many components of b as possible equal to zero and thereby suppressing the recursion for the related branch. This could be done by shifting the most degenerate vertex of the face under consideration into the origin. In practice, however, we do not want to use information about vertices in order to preserve a pure $\mathcal{H}$-based algorithm. Thus we shift a face in dimension k only into a basis solution by solving an arbitrary, non-degenerate $k \times k$ subsystem of equations. In the case of hypercubes this decreases the number of constraints with non-zero b-component in the reduced systems of dimension k from $2k$ to k. The resulting algorithm, including parallel face detection and shifting, is called *LAS*.

Proposition 2. *The time complexity of LAS applied to a d-hypercube is $O(d!)$.*

Proof. A similar analysis as in Proposition 1 can be made. To pass from e^k to e^{k-1} we choose among k faces with non-zero b-component, resulting in the recursive equation $s_{k-1} = ks_k$. With $s_d = 1$ it follows that $s_k = \frac{d!}{k!}$. Fixing one constraint together with shifting requires $O(k^3)$ operations for solving a system of linear equations, and the subsequent check for multiple restrictions (which detects the hyperplane parallel to the newly fixed one) needs $O(k^3)$ steps, yielding a complexity bounded by $O(\sum_{k=1}^{d} s_k k k^3) = O(d! \sum_{k=1}^{d} \frac{k^3}{(k-1)!})$. Now $\sum_{k=1}^{d} \frac{k^3}{(k-1)!}$ is bounded by $22.5 + 11 \sum_{k=0}^{d-4} \frac{1}{k!}$, and with $\sum_{k=0}^{d} \frac{1}{k!} < e$ we derive a total complexity bound of $O(d!)$. $\qquad\square$

Note that no extra determinant computation is needed because this is implicitly done in the course of fixing constraints. LAS improves on C&H by a factor of d^3.

LAWRENCE'S VOLUME FORMULA. Assume P is *simple* and choose a vector $c \in \mathbb{R}^d$ and a scalar q such that the function $x \mapsto c^T x + q$ is not constant along any edge. For each vertex $v \in V$ let A_v be the $d \times d$-matrix composed by the rows of A which are binding at v. Then A_v is invertible and $\gamma^v := (A_v^T)^{-1} c$ is well defined up to permutations of the entries. The assumption imposed on c assures that none of the entries of γ^v is zero. Lawrence [15] shows that then

$$\text{Vol}(P) = \sum_{v \in V} \frac{(c^T v + q)^d}{d! \, |\det A_v| \, \prod_{i=1}^{d} \gamma_i^v}. \tag{6}$$

Lawrence's formula easily generalises to the non-simple case using standard lexicographic perturbation techniques. Consequently the summation in (6) must be done over all lexicographically feasible cobases of all $v \in V$.

An open question is how to choose c and q properly; even if $c^T x + q$ is not constant on any edge, a 'nearly constant' choice results in very small entries γ_i^v in the denominator, causing potentially severe numerical problems. Indeed, such numerical instabilities were regularly found in our experiments. It should be noted that a theoretically satisfactory[2] choice is given in [15], but its practical importance is probably very small since the number of digits needed to represent a theoretically guaranteed vector will be too long to perform any reasonable computation.

The time complexity for the simple case is $O(d^3 n)$, where n denotes the number of vertices, and for cubes this implies a complexity of $O(d^3 2^d)$.

4. Improvements and a hybrid method

The algorithms described in the previous section can be improved both theoretically and practically by a few simple ideas. In this section we describe such improvements for C&H and LAS. Moreover we present a hybrid approach called HOT for 'hybrid orthonormalisation technique'. To underline the qualitative differences of the methods a few complexity estimates are given.

4.1. Modified Cohen & Hickey's method (C&H)

The major advantage of a triangulation method using both the $\mathcal{H}$- and $\mathcal{V}$-representation is the possibility for a computationally cheap, but powerful test on empty or simplicial faces. Consider in the recursive process a face e^k; then we know that the volume of e^k in the appropriate k-dimensional affine embedding is zero if the number of vertices in e^k is smaller than $k+1$. And if the number of vertices is exactly $k+1$ the recursive process can stop with a face which is either a k-simplex or has volume zero. Even though this does not change the behaviour on simple polytopes like hypercubes, i.e. Proposition 1 still applies, simplicial polytopes have a reduced complexity of $O(md^3)$, where m denotes the number of facets.

Another improvement uses degeneracy information: If the vertices are sorted decreasingly by the number of incident hyperplanes, the number of e^{k-1}'s not containing v_k may drop significantly.

We abbreviate the algorithm obtained by incorporating these modifications again by C&H.

4.2. Revised Lasserre's method (r-LAS)

Observing that when the same constraints are fixed in a different order, the same face may appear several times in (4), it is natural for Lasserre's method to store and subsequently reuse the (projected) face volumes; storing can be done, e.g., in a balanced tree. We call the resulting algorithm r-LAS. Due to the projection, not only the indices of the fixed constraints I characterising the face e^k have to be stored, but also the fixed variables $\bar{J}$ representing the projection space $\Pi_J = \mathbb{R}^d \cap \{x \mid x_j = 0 \ \forall j \in \bar{J}\}$, where $J = \{1, \ldots, d\} \setminus \bar{J}$ denotes the set of free variables. Once the volume of a face e^k projected onto Π_J is known, its projection onto $\Pi_{J'}$, where $J \neq J'$ but $|J| = |J'|$, can be found directly by computing the determinant of the transformation $\Pi_J \rightarrow \Pi_{J'}$. Denote by A_{IJ} the matrix consisting of the rows and columns of A indexed by I and J respectively. In a first step, we 'deproject' $\Pi_J \rightarrow$

---

[2]It is possible to select the vectors c and q whose binary lengths are polynomially bounded by the input size.

$\mathbb{R}^d$, $x_J \mapsto x = \begin{bmatrix} x_J \\ x_{\bar{J}} \end{bmatrix}$, with $x_{\bar{J}} = A_{I\bar{J}}^{-1}(b_I - A_{IJ}x_J)$. In a second step x is projected on $\Pi_{J'}$, $x \mapsto x_{J'}$, by dropping the components related to $\bar{J}'$. Therefore the rows J' of the linear part of the map $x_J \mapsto x$ yield a nonsingular $(d-k) \times (d-k)$ matrix whose determinant s fixes the projective effect, i.e. $\text{Vol}(\Pi_{J'}(e^k)) = |s| \cdot \text{Vol}(\Pi_J(e^k))$.

Proposition 3. *The time complexity of r-LAS applied to a d-hypercube is $O(d^4 3^d)$.*

Proof. (Cf. the proof of Proposition 2) Again, let s_k denote the number of faces considered in dimension k. Starting with any of the k-dimensional faces, we have to fix k constraints and subsequently check for identical and parallel constraints. Each of these steps can be done in time $O(k^3)$. We obtain ks_k faces of dimension $k-1$. Each of these faces has to be looked up in the balanced tree containing the projected face volumes. Taking into account that at most all 3^d faces of the cube are stored in the tree, and that one key comparison can be done with $O(m-k) \subseteq O(d)$ index comparisons, the effort for one look-up is in $O(d^2)$. Notice now that some of the ks_k faces are found in the tree, causing each an overhead of $O(d^3)$ for the deprojection. So the total complexity is in $O(ks_k(k^2 + d^2 + d^3)) \subseteq O(d^4 s_k)$. We now claim that s_k, the number of k-dimensional faces stored in the tree, is bounded above by the total number $2^{d-k}\binom{d}{k}$ of k-dimensional faces. This is true since in a hypercube, which is simple, each non-empty k-dimensional face is the intersection of a unique set of $d-k$ facets. Moreover, empty faces are not stored, for this would involve an intersection of parallel hyperplanes. Hence the overall arithmetic complexity is $O(d^4 \sum_{k=0}^{d} 2^{d-k}\binom{d}{k}) = O(d^4 3^d)$. $\qquad\square$

4.3. A hybrid method (HOT)

In the course of our experiments we identified two critical factors explaining an essential part of the differences in computation time if Cohen & Hickey's triangulation method is compared with Lasserre's signed decomposition method: Empty face detection and storing/reusing intermediate results.

Triangulation methods using vertex and incidence information permit an effective test on simplicial or empty faces by simply counting the number of vertices. Signed decomposition methods like Lasserre's one are usually facet-based where a test on empty faces, e.g. by means of linear programming, is significantly more costly. The situation is reversed when storing/reusing intermediate results is considered. Here triangulation methods like Cohen & Hickey's algorithm suffer from the fact that they are based on simplex volumes and not on face volumes. Lasserre's scheme on the other hand works directly with the (projected) volumes of faces, offering a simple way of storing and reusing face volumes in the recursion process.

Among the possibilities to construct a hybrid method, one can include an empty face detection in Lasserre's method or a storing/reusing scheme in Cohen & Hickey's method. Concerning the first possibility we tested an LP-based empty face detection in Lasserre's algorithm, but it turned out to be not competitive in most cases.[3] On the other hand, storing the whole sub-triangulation of a face in Cohen & Hickey's algorithm would require by far too much memory.

---

[3] Our tests are based on the efficient callable library of cplex using restarting techniques.

So we adopted a different approach, which nevertheless is based on Cohen & Hickey's enumeration of faces and profits from all the improvements described in Section 4.1. Notice that the volume of P can be derived directly from the volume of its facets by the formula

$$\mathrm{Vol}_d(e^d) = \frac{1}{d} \sum_{e^{d-1}:v_d \notin e^{d-1}} \mathrm{Vol}_{d-1}(e^{d-1}) \, \mathrm{dist}(v_d, \mathrm{aff}(e^{d-1})), \qquad (7)$$

where Vol_d denotes the d-dimensional volume, aff the affine hull of a set, and $v_d = \eta(e^d)$. This corresponds to a subdivision of P into pyramids with apex v_d and bases e^{d-1} instead of simplices. The distance between v_d and $\mathrm{aff}(e^{d-1})$ is easily computed if an orthonormal basis of the affine subspace spanned by e^{d-1} is known, which can be computed using Householder transformations. The face volumes $\mathrm{Vol}_k(e^k)$ are stored in a balanced tree with the vertices as key; they are reused whenever the same face reappears during the recursion. We call the resulting algorithm *hybrid orthonormalisation technique* (HOT). Since storing the orthonormal bases would require an enormous amount of memory, they are recomputed from scratch when a $\mathrm{Vol}_k(e^k)$ is reused.

Proposition 4. *The time complexity of HOT applied to a d-hypercube is $O(d^2 4^d)$.*

Proof. (Cf. the proof of Proposition 3) In dimension k, a number of s_k faces are intersected by k facets, and each of the resulting faces is then looked up in the tree, requiring $O(k\, s_k\, 2^k d)$ comparisons, cf. the related complexity analysis for C&H in Proposition 1. By s_{k-1} we denote the number of $(k-1)$-dimensional faces where the recursion proceeds because the corresponding volume is not yet known. After these recursions are finished, $O(d^2 s_{k-1})$ steps are needed for adding one vector to the orthonormal basis. For the remaining $ks_k - s_{k-1}$ faces, which are retrieved from the tree, orthonormal bases are computed from scratch, taking $O(d^3(ks_k - s_{k-1}))$ steps. Adding together, the total complexity is bounded by $O(\sum_{k=1}^d (ks_k 2^k d + d^2 s_{k-1} + d^3(ks_k - s_{k-1}))) \subset O(d^2 \sum_{k=0}^d s_k 2^k + d^4 \sum_{k=0}^d s_k)$. To estimate s_k note that a face is uniquely determined by the set of its vertices, so the same face can be stored only once in the tree. As before, it follows that $s_k \leq 2^{d-k} \binom{d}{k}$. We obtain therefore an overall arithmetic complexity of $O(d^2 2^d \sum_{k=0}^d \binom{d}{k} + d^4 \sum_{k=0}^d 2^{d-k} \binom{d}{k}) = O(d^2 4^d + d^4 3^d)$. $\qquad\square$

Note that even though both r-LAS and HOT store and reuse face volumes, the former does it as projected volumes and the latter as the original (absolute) volumes. As a consequence, one has to deal either with deprojections or the computation of bases (by orthonormalisation). In principle one could also set up a Lasserre scheme based on unprojected volumes. While this would not change the essential behaviour of Lasserre's algorithm, it might be interesting to test its practicality. We leave this as a future research project.

5. Experimental results

The implemented algorithms have been tested on a broad range of examples, varying from polytopes created with random vertices or hyperplanes to specially structured polytopes as they arise in combinatorics. We denote by d the dimension, by m the number of hyperplanes and by n the number of vertices. The groups of examples are ordered from large to small ratio n/m:

- cube-d: Hypercube in dimension d.
- rh-d-m: d-dimensional problems with m randomly generated constraints. The normal vectors of the constraints have integer components and a vector norm of approximately 1000; the right hand side is fixed to 1000. The resulting polytopes are (most probably) simple.
- $CC_d(k)$: The product of two cyclic polytopes over k points in dimension d.
- Fm-d: One facet of the metric polytope in dimension d.
- ccp-v: Complete cut polytope on v vertices, scaled by two and translated by one.
- rv-d-n: d-dimensional convex hulls of n randomly generated vertices on a sphere with radius 1000. The resulting polytopes are (most probably) simplicial.
- cross-d: Cross polytope of dimension d; it is the polar of the d-dimensional cube.

A discussion of these examples can be found in [1]. We emphasise that our main concern is the comparison of the general behaviour of the methods and not of specific implementations; furthermore, most of the codes are experimental in nature leaving room for improvements. The observed time should therefore be interpreted cautiously.

Unless otherwise stated the computed volume is identical for all methods for at least six digits and is therefore given only once. The general abbreviations for the codes used in the sequel are given below in the first column; in the second column the shorter ones used in Table 3 and 4 are stated.

BND	Bnd	boundary triangulation using 'lrs'
DEL	Del	Delaunay triangulation using 'qhull'
C&H	CH	Cohen & Hickey's triangulation including our improvements
HOT	HOT	hybrid orthonormalisation technique
LAW-nd	Lnd	Lawrence's formula in the non-degenerate case
LAW-d	Ld	Lawrence's formula in the general case using 'lrs' for computing all lexicographically feasible cobases
LAS		original Lasserre
r-LAS	rL	revised Lasserre

Sources for the codes are given in the appendix. The only code using rational arithmetic is 'lrs'. All computations have been done on an HP 7000/735–99. Among the many parameters influencing the behaviour of the codes the one determining the level of storing face volumes is of major importance in case of HOT and r-LAS. For HOT and most of the examples computed with r-LAS it was possible to store all faces; in the remaining examples the choice was done to take most advantage of the memory available which was limited to approximately 500 MB. Obviously, for larger problems there is a decisive trade-off between memory and CPU usage if face volumes are stored/reused. In case of r-LAS another influencing parameter is the minimal absolute value MIN_PIVOT accepted for pivoting; the strategy is to go through a row and pick the first element which exceeds MIN_PIVOT in absolute value. If none is found the absolutely largest element is chosen. The larger MIN_PIVOT, the slower r-LAS because the probability of projecting the same face onto the same subspace decreases. In our experiments we set MIN_PIVOT to 0.01. A too small MIN_PIVOT, however, contains the risk of numerical instabilities.

In the following two sections we examine the effect of our modifications on Cohen & Hickey's and Lasserre's algorithms. Then a comparison of the modified methods with all other codes on the full set of test problems is presented. Here we drop the original C&H and LAS because they are not competitive.

5.1. C&H and its modified version

As described in the previous section, the original Cohen & Hickey triangulation has been improved in two ways: Detection of simplicial faces leads to an early abort strategy, and ordering the vertices following the number of incident hyperplanes potentially decreases the number of computed simplices. Table 1 reports the observed running time for both the original and the improved code on cubes and cross polytopes.

Problem	d/m/n	original C&H	modified C&H
cube-7	7/14/124	1.5	1.0
cube-8	8/16/256	13	7.6
cube-9	9/18/512	130	82
cube-10	10/20/1024	1400	940
cube-11	11/22/2048	18000	14000
cross-7	7/124/14	0.3	0.1
cross-8	8/256/16	0.8	0.2
cross-9	9/512/18	3.4	0.3
cross-10	10/1024/20	15	0.6
cross-11	11/2048/22	71	1.4
cross-12	12/4096/24	340	4.3
cross-13	13/8192/26	1700	15
cross-14	14/16384/28	7600	65

TABLE 1. The original and the modified version of C&H

Since both cubes and cross polytopes have a constant number of incident hyperplanes for all vertices, vertex reordering has no effect. Concerning the detection of simplicial faces, cubes profit only a little; because they are simple the recursion still has to go down to the edges.

The situation changes dramatically for cross polytopes. Since they are simplicial, the recursive scheme is reduced to computing a determinant for half of the simplicial facets.

As an example where vertex reordering has a significant impact, we observed in case of Fm_6 a drop from 8400 CPU-seconds (reversed ordering) to 25 s. (ordered following maximal incidence), where in both cases simplicial face detection was enabled. Fm_6 also demonstrates the practical gain of our early abort strategy for neither simple nor simplicial polytopes. Using the optimal vertex ordering, the running time rises from 25 s to 4600 s if the detection of simplicial faces is turned off.

5.2. LAS versus r-LAS

In Table 2 the effects of shifting into a basis solution and storing face volumes are demonstrated for cubes and cross polytopes. The first component in each pair of signs relates to shifting, the second component to storing. A '+' means active, a '−' means not enabled. The column $(+, -)$ represents Lasserre's original algorithm, $(+, +)$ stands for our revised code r-LAS.

Problem	d/m/n	Variants of Lasserre			
		$(-, -)$	$(+, -)$	$(-, +)$	$(+, +)$
cube-7	7/14/124	4.4	0.13	0.26	0.0
cube-8	8/16/256	69	1.4	1.4	0.0
cube-9	9/18/512	1200	13	6.3	0.13
cube-10	10/20/1024	25000	130	21	0.26
cross-7	7/124/14	22	11		
cross-8	8/256/16	340	180		
cross-9	9/512/18	5400	2900		
cross-10	10/1024/20	92000	48000		

TABLE 2. Lasserre's algorithm with (+) or without (−) shifting and storing, respectively. The case $(+, -)$ represents LAS, whereas $(+, +)$ is r-LAS.

For cubes the effects of these two factors reduce the CPU-time drastically, where the effect of storing is slightly more important than that of shifting. In case of cross polytopes, however, the facet-based approach of r-LAS has severe difficulties exploiting the structure. First of all, storing has no effect because no face ever appears twice, thus $(?, -)$ equals $(?, +)$, for $? \in \{+, -\}$. Secondly, shifting is only done once at the very beginning because in every recursion level $k > 0$, already at least $d - k$ components of the right hand side are zero. Therefore we observe a CPU-time without shifting which is roughly twice the time needed when shifting is active.

5.3. Comparing the main codes

In Table 3 the timing in CPU-seconds for all codes and examples is given; because the transformation $\mathcal{V}\to\mathcal{H}$ or $\mathcal{H}\to\mathcal{V}$ can be as demanding as the volume computation itself, and furthermore we need both representations for certain methods, we give in the last two columns the transformation time when using 'lrs' and 'pd' with exact arithmetic, see Appendix 6.

| | | | | | | Triangulation | | | | Signed decmp. | | | | |
| | | | | | | $\mathcal{V}$-r. | | $\mathcal{V}$- and $\mathcal{H}$-r. | | $\mathcal{H}$-r. | | | | |
Exp.	$\dfrac{n}{m}$	d	m	n	vol.	Bnd	Del	CH	HOT	Lnd	Ld	rL	$\mathcal{H}\to\mathcal{V}$	$\mathcal{V}\to\mathcal{H}$
cube-9	28.4	9	18	512	512	25e3	270	82	4.0	.3	1.6	.3	1.1	12
cube-10	51.2	10	20	1024	1024	$**^c$	$**^a$	940	18	.6	3.3	.5	2.7	36
cube-14	585	14	28	16384	16384	$**^c$	$**^a$	$**^c$	3300	13	90	10	88	3500
rh-8-20	56	8	20	1115	37576	$**^c$	$**^b$	93	7.8	1.5	28	14	28	1900
rh-8-25	104	8	25	2596	786.0	$**^c$	$**^b$	430	27	3.5	80	120	80	4600
rh-10-20	109	10	20	2180	13883	$**^c$	$**^b$	3000	63	5.2	90	31	91	9500
rh-10-25	309	10	25	7724	5729.5	$**^c$	$**^b$	34e3	390	15	390	600	380	44e3
$CC_8(9)$	1.5	8	54	81	13340	410	40	11	3.4	$**^c$	75	140	72^d	140^d
$CC_8(10)$	1.43	8	70	100	156816	1300	80	28	7.7	$**^c$	210	880	200^d	340^d
$CC_8(11)$	1.38	8	88	121	1.39e6	3500	150	65	17	$**^c$	550	4400	550^d	800^d
Fm-6	3	15	59	177	286114	11e5	$**^a$	25	12	$**^c$	$**^b$	6900^9	$13e3^d$	6600
ccp-5	0.3	10	56	16	2.3117	.7	.5	.2	.2	$**^c$	200	2900^6	4.5	.7
ccp-6	0.09	15	368	32	1.3458	520	150	43	23	$**^c$	$**^c$	$**^c$	4400^d	2000^d
rv-8-10	0.42	8	24	10	1.41e19	.3	.2	.1	.1	$**^c$	82	.2	16	.5
rv-8-11	0.2	8	54	11	3.05e18	.4	.1	.2	.1	$**^c$	2100	79	33	.9
rv-8-30	0.007	8	4482	30	7.35e21	170	4.0	6.9	6.8	$**^c$	$**^c$	$**^c$	7200	150
rv-10-12	0.34	10	35	12	2.14e22	.4	.2	.1	.2	$**^c$	1300	.3	92	1.0
rv-10-14	0.08	10	177	14	2.93e23	1.9	.2	.2	.1	$**^c$	$**^c$	$**^c$	510	5.1
cross-8	0.063	8	256	16	6.35e-3	.5	.3	.2	.2	$**^c$	2800	170	4.2	.5
cross-9	0.035	9	512	18	1.41e-3	0.9	.5	.3	.3	$**^c$	$**^b$	2700	12	1.2

TABLE 3. Timing for the different methods and examples.
[a] beyond memory limit, [b] volume computed incorrectly,
[c] problem is intractable with this method, [d] 'cdd' is faster by a factor of at least 100, [6,9] storage performed for 6 and 9 levels resp.

Next, to measure potential numerical difficulties, Table 4 presents condition numbers for some algorithms. As different methods face different numerical problems we decided to define two different condition numbers: For triangulation methods all partial volumes are positive; the only problem lies in summing up volumes of potentially different magnitudes. Hence in this case we define the condition number as $\log_{10}$ of the biggest simplex volume divided through the smallest

nonzero one (zero volumes occur with 'qhull', which returns some degenerate simplices). When evaluating Lawrence's formula the main problem is that positive
and negative numbers have to be summed up whose absolute values may be much
larger than the final result. Here we use as condition number the biggest absolute
summand divided through the correct volume, again as logarithm to the base 10.
This number would be at most zero for triangulations. All condition numbers are
rounded up.

Example	Del	CH	Ld
cube-9	2	0	8
cube-10	**[a]	0	9
cube-14	**[a]	**[c]	14
rh-8-20	**[b]	16	7
rh-8-25	**[b]	15	10
rh-10-20	**[b]	17	13
rh-10-25	**[b]	16	12
$CC_8(9)$	6	6	10
$CC_8(10)$	7	7	8
$CC_8(11)$	8	8	8
Fm-6	**[a]	3	**[b]
ccp-5	1	1	11
ccp-6	2	1	**[c]
rv-8-10	1	1	7
rv-8-11	3	2	11
rv-8-30	6	6	**[c]
rv-10-12	1	1	10
rv-10-14	4	3	**[c]
cross-8	0	0	11
cross-9	0	0	**[b]

TABLE 4. Condition numbers for the different methods and examples. [a]beyond memory limit, [b]volume computed incorrectly,
[c]problem is intractable with this method.

Based on the Tables 3 and 4 we comment on the efficiency and the numerical
stability of each algorithm separately. As a general remark we find huge differences
in CPU time among the different codes for the same polytope. Some problems are
even intractable with certain methods whereas they are quite efficiently solved by
others, demonstrating the relevance of a good choice. Moreover, we regularly faced
practical problems ranging from insufficient memory to numerical instabilities.
    The boundary triangulation BND, done by 'lrs' using rational (exact) arith-
metic, requires in general more CPU-time, but can nevertheless be competitive

for some near-simplicial polytopes with small ratio n/m. In all tested cases this method produced the largest number of simplices compared to the other triangulation methods.

The Delaunay triangulation DEL produced by 'qhull' exhibits numerical problems in many examples. For specially structured polytopes, where the vertices are not in general position, numerical perturbation is needed, which may cause a break-down as observed with hypercubes. Moreover, the memory space needed for the quick hull algorithm and in fact all incremental algorithms, see [2], is not bounded above by a polynomial in the input and output size, and quite often a memory overflow occurs. The condition number is comparable to C&H, as well as the number of generated simplices.

C&H turns out to be numerically very robust due to its combinatorial nature. For special problems it profits a lot from degeneracy information, e.g. for Fm-6 where one vertex lies in all but one facet. (This information is not used by r-LAS.) The bigger the ratio n/m, however, the more C&H is slowed down by the increasing number of simplices.

The new hybrid code, HOT, is uniformly faster than C&H. Compared to r-LAS the situation is slightly different. Leaving apart hypercubes, the hybrid method is faster in most cases, but suffers from the burden of computing orthonormal bases for retrieved faces. Considering the memory requirement, HOT is in most problems more modest than r-LAS. To explain the tremendous effect of reusing face volumes in HOT, we observed for increasing $\frac{n}{m}$ an increased average reuse of already computed face volumes. Thus HOT handles just the examples efficiently where C&H fails while making still use of C&H's strong point in the other examples, namely its detection of simplicial faces.

Numerical instabilities, partly due to the unresolved problem of choosing the objective function, are a serious drawback of Lawrence's approach presented in the columns labelled LAW-nd (non-degenerate) and LAW-d (degenerate: 'lrs' is used for finding all cobases). However, for most of the examples the result is correct, even though in some cases positive and negative summands occurred with absolute value going by a factor of 10^{14} beyond the final volume. In such a situation numerical cancellation can drastically destroy the accuracy of the final result. For simple polytopes LAW-nd is the fastest method exploiting efficiently the double description.

r-LAS is quite efficient for near-simple polytopes with a high ratio $\frac{n}{m}$. It profits a lot from memory for storing intermediate results. Due to the detection of parallel facets it works extremely fast on hypercubes. If the number of hyperplanes increases, however, time and/or memory requirement can grow exceedingly large, see ccp-6, rv-8-30 or rv-10-14. Numerically r-LAS performs almost as robustly as C&H in our examples.

The transformation times between $\mathcal{H}$- and $\mathcal{V}$-representations, given in the last two columns, have to be interpreted with extra caution. Most of the known algorithms face an 'easy' and a 'difficult' direction, where, roughly speaking, the easy transformation is the one from few vertices or facets to many facets or vertices, and the difficult situation is the converse direction. (For a thorough discussion of these issues, see [1].) We observed that especially the more demanding transformation can be several times more time consuming than the volume computation itself. A new result, however, changes this situation: It shows that using the easy direction as an oracle to the hard one, this latter task is considerably simplified [3]. Up to date, only one implementation of these so-called primal-dual algorithms is available, namely 'pd', which uses reverse search with exact arithmetic as the oracle. In Table 3 the transformation times using 'lrs' for the 'easy' and 'pd' for the 'difficult' direction are reported. Still, the transformations can be more time consuming than the actual volume computation, so it seems worthwhile to take the representation into account when choosing an algorithm for the volume computation. We hope, however, that such primal-dual approaches, especially when using floating point arithmetic, will considerably ease the burden of transformation in the near future. It should also be noted that some hard transformations, like $\mathcal{V}$- to $\mathcal{H}$-representation for random hyperplanes, are very unlikely to occur in practice.

6. Conclusions

The experimental results indicate that vertex based triangulations are superior for a small ratio n/m, namely for simplicial polytopes, whereas constraint based signed decomposition codes behave favourably for large n/m, e.g. for simple polytopes. Outstanding is – except for hypercubes – the behaviour of the hybrid code HOT; it convinces on both near-simple and near-simplicial polytopes, and exhibits a modest need for memory on our set of examples. Thus, in the general case, given both representations or an efficient transformation code, HOT seems to be the method of choice.

While we investigated many of the existing techniques for computing the volume of polytopes, this research can be considered as a starting point. For example, it might be interesting to study how a specific volume computation algorithm can be extended to compute mixed-volumes (see, e.g. [17]) and to investigate their complexities. Another natural extension is to study the efficient integral computation over a polytope, cf. [14]. We leave these for future research.

Appendix A: Availability of the codes

All algorithms described in Section 3 and 4 have been implemented in C by the authors, using four publicly available additional programs, namely 'cdd' (double description in C, transfers between $\mathcal{H}$- and $\mathcal{V}$-representations), 'qhull' (quick hull, computes convex hulls for a given set of vertices), 'lrs' (lexicographic reverse search, constructs convex hulls and cobases using lexicographic perturbation), and 'pd' (primal-dual version of reverse search). The resulting volume computation package called 'vinci' provides a common framework for the different methods and codes.

While C&H, HOT and r-LAS are implemented directly in 'vinci', the remaining algorithms – Delaunay triangulation, boundary triangulation and Lawrence's method – are realised using 'qhull' for the first and 'lrs' for the other two methods; in the special case of a simple polytope the incidence information suffices to generate Lawrence's signed decomposition and hence a direct summation, omitting 'lrs', can be performed. The necessary communication is done via text files. 'Vinci' checks the input and calls the appropriate external code if necessary. After the termination of 'qhull' or 'lrs', 'vinci' reads the resulting triangulation or signed decomposition and makes the appropriate summation of simplex volumes. Concerning Lawrence's method we use random values for c and q. A thorough discussion of 'cdd', 'qhull' and 'lrs' can be found in [1, 3], the algorithm implemented in 'pd' is described in [3]. 'Vinci' and the additional programs are freely available at the sites given in Table 5.

Acknowledgement. We are grateful to Jean-Bernard Lasserre for encouragement and fruitful discussions related to his volume computation method, and to Paul A. Vixie[4] for giving us his efficient avl-tree implementation. We thank Günter Rote for fruitful discussions, and Gyula Karolyi for pointing out Filliman's paper [8] to us, which turned out to be very useful in classifying algorithms.

References

[1] D. Avis, D. Bremner, and R. Seidel. How good are convex hull algorithms. *Computational Geometry: Theory and Applications*, 7:265–302, 1997.

[2] D. Bremner. Incremental convex hull algorithms are not output sensitive. *Discrete Comput. Geom.* **21** (1999), 57–68.

[3] D. Bremner, K. Fukuda, and A. Marzetta. Primal-dual methods for vertex and facet enumeration. In *Proc. 13th Annu. ACM Sympos. Comput. Geom.*, pages 49–56, 1997. Full paper in *Discrete Comput. Geom.* **20** (1998), 333–357.

[4] J. Cohen and T. Hickey. Two algorithms for determining volumes of convex polyhedra. *Journal of the ACM*, 26(3):401–414, July 1979.

[5] M. E. Dyer and A. M. Frieze. The complexity of computing the volume of a polyhedron. *SIAM J. Comput.*, 17:967–974, 1988.

[6] M.E. Dyer. The complexity of vertex enumeration methods. *Math. Oper. Res.*, 8:381–402, 1983.

[7] H. Edelsbrunner. *Algorithms in Combinatorial Geometry*. Springer-Verlag, 1987.

[8] P. Filliman. The volume of duals and sections of polytopes. *Mathematika*, 39:67–80, 1992.

[9] K. Fukuda, T. M. Liebling, and F. Margot. Analysis of backtrack algorithms for listing all vertices and all faces of a convex polyhedron. *Computational Geometry*, 8:1–12, 1997.

---

[4]Internet Software Consortium, Star Route Box 159A, Woodside, CA 94062 USA, vixie@vix.com, http://www.isc.org/isc/

vinci	authors:	Benno Büeler (bueeler@ifor.math.ethz.ch) and Andreas Enge (enge@ifor.math.ethz.ch)
	www page:	http://www.mathpool.uni-augsburg.de/~enge
	ftp site:	ftp.ifor.math.ethz.ch (129.132.154.13)
	directory:	pub/volume
	file name:	vinci-*.tar.gz, where '*' stands for the actual version number
lrs	author:	David Avis (avis@cs.mcgill.ca)
	ftp site:	mutt.cs.mcgill.ca (132.206.3.13)
	directory:	pub/C
	file name:	lrs*.c, where '*' stands for the actual version number
pd	author:	Ambros Marzetta (marzetta@inf.ethz.ch)
	www page:	http://wwwjn.inf.ethz.ch/ambros/pd.html
qhull	authors:	Brad Barber (bradb@geom.umn.edu) and Hannu Huhdanpaa (hannu@geom.umn.edu)
	www page:	http://www.geom.umn.edu/locate/qhull
	ftp site:	ftp.geom.umn.edu (128.101.25.35)
	directory:	pub/software
	file name:	qhull-*.*.tar.Z, where '*.*' stands for the actual version number
cdd+	author:	Komei Fukuda (fukuda@ifor.math.ethz.ch)
	www page:	http://www.ifor.math.ethz.ch/ifor/ staff/fukuda/cdd_home/cdd.html
	ftp site:	ftp.ifor.math.ethz.ch (129.132.154.13)
	directory:	pub/fukuda/cdd
	file name:	cdd+-***.tar.gz, where '***' stands for the actual version number

TABLE 5

[10] K. Fukuda and A. Prodon. Double description method revisited. In M. Deza, R. Euler, and I. Manoussakis, editors, *Combinatorics and Computer Science*, volume 1120 of *Lecture Notes in Computer Science*, pages 91–111. Springer-Verlag, 1996.

[11] P. Gritzmann and V. Klee. On the complexity of some basic problems in computational convexity: II. Volume and mixed volumes. In T. Bisztriczky, P. McMullen, R. Schneider, and A.I. Weiss, editors, *Polytopes: Abstract, convex and computational (Scarborough, ON, 1993)*, NATO Adv. Sci. Inst. Ser. C Math. Phys. Sci., 440, pages 373–466. Kluwer Acad. Publ., Dordrecht, 1994.

[12] R. Kannan, L. Lovász, and M. Simonovits. Random walks and an $O^*(n^5)$ volume algorithm for convex bodies. *Random Struct. Algorithms*, 11(1):1–50, 1997.

[13] J. B. Lasserre. An analytical expression and an algorithm for the volume of a convex polyhedron in $\mathbb{R}^n$. *J. of Optimization Theory and Applications*, 39(3):363–377, 1983.

[14] J. B. Lasserre. Integration on a convex polytope. Technical Report 96173, Centre National de la Recherche Scientifique, Laboratoire d'Analyse et d'Architectures des Systems, Toulouse, 1996.

[15] J. Lawrence. Polytope volume computation. *Mathematics of Computation*, 57(195):259–271, 1991.

[16] R. Seidel. Small-dimensional linear programming and convex hulls made easy. *Discrete Comput. Geom.*, 6:423–434, 1991.

[17] J. Verschelde, K. Gatermann, and R. Cools. Mixed-volume computation by dynamic lifting applied to polynomial system solving. *Discrete Comput. Geom.*, 16:69–112, 1996.

Benno Büeler
Institute for Operations Research
Swiss Federal Institute of Technology
CH-8092 Zurich, Switzerland

Andreas Enge
Lehrstuhl für Diskrete Mathematik
Optimierung und Operations Research
Institut für Mathematik
Universität Augsburg
Deutschland

Komei Fukuda
Department of Mathematics
Swiss Federal Institute of Technology
CH-1015 Lausanne
Switzerland

Reconstructing a Simple Polytope from its Graph

Hans Achatz and Peter Kleinschmidt

Abstract. Let P be a simple polytope with dimension d and $G(P)$ its edge graph. It has been shown in [BlM87] and [Kal88] that $G(P)$ already determines the complete face-lattice of P. However, the constructive approach used in [Kal88] requires the computation of all orderings in $|vert(P)|$ which is computationally prohibitive for polytopes of even very small sizes. In this paper we propose an algorithm which is still exponential but does work with reasonable computing time for non-trivial simple polytopes.

1. Introduction

Throughout this paper P will be a simple polytope and $G(P) = (V, E)$ its edge graph. $G(P)$ is known to be d-regular and d-connected.

The following result was conjectured by Perles and first proved by Blind and Mani-Levitska.

Theorem 1 (Blind, Mani-Levitska [BlM87]).
The face-lattice of P is determined by $G(P)$.

Theorem 1 is not correct for non-simple polytopes.

Kalai [Kal88] gave a simple proof for Theorem 1. His construction is based on the concept of an abstract objective function.

Definition 1. *An **abstract objective function (AOF)** is an ordering of vertices of P such that for every nonempty face P has a unique local maximum vertex w.r.t. the ordering.*

The following lemma [Kal88] is crucial for the construction of all faces of P.

Lemma 1. *It is possible to determine whether an ordering of V is an AOF.*

We need the following argument for the proof of Lemma 1. Let $<$ be an ordering of V. For $v \in V$ let $deg(v)$ denote the number of vertices adjacent to v which are smaller than v with respect to $<$. Let $h_k^<$ denote the number of vertices of degree k, $0 \le k \le d$. The ordering $<$ is an AOF if and only if the quantitiy

Supported by GIF Grant I-0309-146.06/93.

$\varphi^< = \sum_{k=0}^d 2^k h_k^<$ is minimum for all orderings of V. Hence for deciding whether an ordering is an AOF based on the approach of the lemma one has to compute this quantity $\varphi^<$ for all orderings of V (or at least for all cycle free orientations of $G(P)$).

The following Theorem 2 of Kalai reduces the computation of the face-lattice of P to the computation of all k-regular connected induced subgraphs of $G(P)$ ($2 \le k \le d-1$) and the computation of all AOF's by way of Lemma 1.

Theorem 2 (Kalai). *A k-regular connected induced subgraph F of $G(P)$ is the graph of a k-face if and only if there exists an AOF such that the vertices of F are smaller than all other vertices of P.*

In Section 2 we will describe an algorithm which constructs the vertex sets of all facets of P from $G(P)$. Based on this knowledge there are various obvious algorithms for determining the complete face-lattice of P. Of course improvements are possible.

Our method certainly does not represent the "ultimate" algorithm for this purpose. The main goal of presenting it is to show that in principle Kalai's approach can be refined to a computationally acceptable algorithm. In Section 3 we present some computational experience for randomly generated polytopes and some polytopes occuring in the literature.

2. The algorithm

In this chapter we will describe an algorithm which computes all facets of a simple d-polytope P whose graph $G(P)$ is given. As candidates for facets we will construct "quasifacets".

Definition 2. *A **quasifacet** of the graph $G(P)$ is a $(d-1)$-regular connected induced subgraph of $G(P)$ which does not separate $G(P)$.*

Of course the graph of every facet of P is a quasifacet. The following conjecture states that the converse is also correct. This conjecture seems to be folklore in convex polytope theory, but we have not found an explicit reference.

Conjecture 1. *A quasifacet of the graph $G(P)$ of a simple d-polytope P is always the graph of some facet of P.*

It will turn out that our algorithm would be much faster if Conjecture 1 had been proved. However, as this is not the case, we have to assume that $G(P)$ contains quasifacets which are not the graph of any facet of P. According to Theorem 2 we can test whether a quasifacet corresponds to some facet or not. Hence determining all quasifacets of $G(P)$ and then excluding the non-facets by use of Theorem 2 solves the problem.

A vertex f in $G(P)$ and any set of $d-1$ edges containing f determine a facet. There is precisely one vertex c adjacent to f which is not in this facet. Therefore, any (ordered) pair (f, c) of adjacent vertices f and c determines precisely one facet

F of P where f is in F and c is not. For such a pair (f, c) let $Q(f, c) = \{S_i | 1 \leq i \leq k\}$ be the set of all vertex sets S_i of quasifacets containing f but not c. We will determine $Q(f, c)$ for all such pairs (f, c). It will turn out that this can be done quite efficiently for polytopes of non-trivial sizes. Only if $|Q(f, c)| > 1$ for some (f, c) we have to determine the unique vertex set in $Q(f, c)$ corresponding to the facet by constructing an AOF as required by Theorem 2. This will be done by solving an integer linear program (ILP) described in Section 2.2. However it turned out that in all our computed examples no such program had to be solved because Conjecture 1 was always valid for them.

In our algorithm, the set $\mathcal{F}$ will store all vertex sets of the so far known facets of P. It is initialized as the empty set. As long as there is a vertex f which has not been decided to lie in exactly d facets of $\mathcal{F}$ there will be some vertex c adjacent to f such that the pair (f,c) determines (in the above mentioned sense) a new facet not in $\mathcal{F}$. For such a pair we compute $Q(f, c)$, determine the unique facet and add the corresponding vertex set to $\mathcal{F}$.

The following pseudo-code describes the structure of the algorithm RECON-STRUCT whose output is the set of all the vertex sets of facets of P stored in $\mathcal{F}$.

ALGORITHM RECONSTRUCT
$\mathcal{F} := \emptyset$
while $\exists f \in V$ such that $|\{F_i | F_i \in \mathcal{F}, f \in F_i\}| < d$
begin
        choose some $c \in V$ adjacent to f such that
         all $F_i \in \mathcal{F}$ containing f also contain c
        compute $Q(f, c)$ /* algorithm AVALANCHE of Section 2.1 */
        if $Q(f, c) = \{S_1\}$
        then $\mathcal{F} := \mathcal{F} \cup \{S_1\})$
        else begin
                compute the unique $S_j \in Q(f, c)$ which is the vertex set
                of a facet /* ILP model of Section 2.2*/
                $\mathcal{F} := \mathcal{F} \cup \{S_j\}$
        end
end.

2.1. Searching for quasifacets

Using the terminology of the last section we will now describe in detail an algorithm AVALANCHE which for a given pair (f_0, c_0) computes the set $Q(f_0, c_0)$. In principle this algorithm is an exhaustive search algorithm with exponential worst case time bound. In order to reduce the number of candidates for the vertices of a particular quasifacet we will use three simple facts on the graph $G(P)$ of a simple d-polytope P:

        Let F be a quasifacet of $G(P)$.

Fact 1: Let v be a vertex of F and u a vertex adjacent to v but not in F then all other vertices of $G(P)$ which are adjacent to v are also vertices of F.

Fact 2: Let u and v be vertices not in F and w a vertex adjacent to both u and v then w is not in F.

Fact 3: Let v be a vertex of F and $u_1, \dots, u_{d-1}$ vertices in F which are adjacent to v then the remaining adjacent vertex of v is not in F.

Given f_0 and c_0, by fact 1 all neighbors of f_0 different from c_0 are in every quasifacet of $Q(f_0, c_0)$. Starting from these vertices we will apply our three facts as long as possible in order to grow the vertex set of an element S of $Q(f_0, c_0)$. If none of the facts is applicable for this purpose we will choose a particular vertex for which the membership in S has not been decided so far. In the procedure `decide` we use some heuristic to decide whether the vertex should be in S or not. After this decision we will proceed with facts as before or apply `decide` again. The decision made in `decide` may be wrong. This is why all alternative decisions will have to be checked also. This results in a decision tree illustrated in Fig. 1.

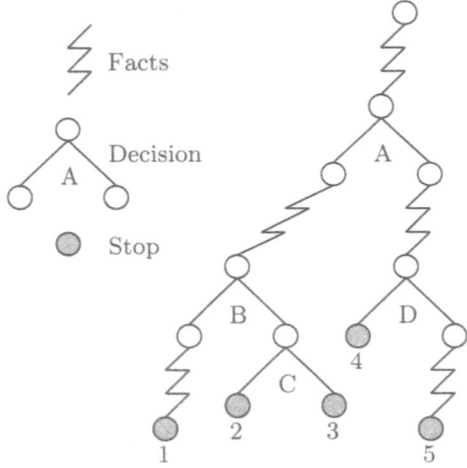

FIGURE 1. Decision tree

As every decision or the use of a fact results in an increase of the number of a candidate set for a quasifacet or its complement, every branch of the decision tree eventually will have to end in some leaf.

In a leaf node the constructed set S must have one of the following properties (which can be tested polynomially):

(i) S corresponds to a quasifacet.
(ii) The graph induced by S is not connected.
(iii) S separates $G(P)$.
(iv) The structure of S is such that because of the decisions made before (in `decide`) there is no way of extending S to the vertex set of a quasifacet.

We will now prove that this way $Q(f_0, c_0)$ is correctly determined by the algorithm. Let us assume that some $S_i \in Q(f_0, c_0)$ is not found by the algorithm.

Before the first call of the procedure decide we have only used facts. This means that every vertex which is decided to be in S is also an element of S_i. When decide has branched the decision tree (w.r.t. vertex u) for the first time, its two branches cover the the case $u \in S$ or $u \notin S$. One of these branches must be correct relative to S_i. The rest of the proof is an obvious recursion of this argument.

It should be noted here that this argument is completely independent of the actual way decide is implemented. This leaves a lot of room for improvements of our algorithm. We implemented decide in a way which turned out to keep the decision tree relatively small.

Apart from using the set S we maintain the following two sets:
C is the set of all vertices which have been decided not to be in S. U is $(V \setminus S) \setminus C$, i.e. the set of vertices whose status w.r.t. S is undecided.

Decide proceeds as follows:

procedure decide()
begin
    if there exists some $v \in U$ which has at least two neighbors in S
    then $S := S \cup \{v\}$, $U := U \setminus \{v\}$ /* s-step */
    else
        if there exist two adjacent vertices u and v in U
            each of which has a neighbor in C
        then $C := C \cup \{v\}$, $U := U \setminus \{v\}$ /* c-step */
        else
            if there exists some $v \in U$ which has a neighbor in C
            then $C := C \cup \{v\}$, $U := U \setminus \{v\}$ /* c-node */
            else stop this branch of the decision tree /* stop */
end;

Observe that the first three cases (s-step, c-step, c-node) of the if-statements use the fact that both graphs induced by S and C have to be connected. The incorporation of more knowledge would most likely improve the algorithm.

It remains to show that the potential last step (stop) in decide in which a branch of the decision tree is terminated in a leaf is correct. This means that we have to show that in this case there exists no quasifacet which contains S and no vertex from C (see case (iv)).

For this proof let S, C and U as before. A node of S is a member of a set $\hat{S}$ if all its neighbors are in $S \cup C$. If S equals $\hat{S}$ then due to our construction we have got a $(d-1)$-regular subgraph induced by S and the branch of the decision tree can be stopped.

Now, suppose that $S \setminus \hat{S}$ is not empty. Let $u \in C$ and $v \notin C$ be adjacent vertices. Then $v \notin U$ because otherwise the case c-node in decide would be applied. Hence $v \in S$. Also, v cannot be in $S \setminus \hat{S}$ because then fact (i) would be applicable and **decide** would not be called here. It follows that $v \in \hat{S}$. This implies that in $G(P)$ the set $\hat{S}$ separates C and U. Let $s \in S \setminus \hat{S}$. Then, we have seen that s has no neighbor in C. Hence there must be exactly one neighbor $u \in U$ of s which cannot become a member of S at a later stage. This implies that u will remain separated from C by S at all time. Hence there is no quasifacet which contains S and no vertex from C.

In our description of the algorithms RECONSTRUCT, AVALANCHE and the procedure **decide** we have not given any details about our actual implementation. Of course there are various choices for implementing the search for vertices which allow the application of facts. The same holds for the obviously necessary back-tracking strategy for handling the decision tree. We decided not to overload this article by implementational details.

The behaviour of the algorithm just described is similar to that of an ava-lanche. The algorithm is triggered by two nodes f_0 and c_0 and by the use of facts is spread over $G(P)$ like an avalanche. If it comes to a halt an artificial decision in **decide** triggers the process again. In our computed examples we observed cases (not many) of non-trivial facets which could be entirely determined by the use of facts, i.e. without the use of **decide**.

If there is more than one element in $Q(f_0, c_0)$ we have to decide which one is the vertex set of the corresponding facet. This is done in the next section.

2.2. Computing AOFs and h-vectors

In this section, we describe how we solve the only missing part of the algorithm RECONSTRUCT. We will determine whether a given quasifacet corresponds to a facet by avoiding the prohibitive computation of all orderings of V. Instead we will use an ILP-model which decides whether an AOF with the properties of Theorem 2 exists. To solve this model we still have to attack a NP-hard problem. However our model is very closely related to the Linear Ordering Problem which has been well studied in Combinatorial Optimization (see e.g. [Rei85]). This allows the use of very sophisticated methods for solving the problem in reasonable time even for larger problems. It should be noted that this part is obsolete if Conjecture 1 is correct.

First we will propose an ILP-model which determines some AOF for $G(P)$. In what follows we will write G for $G(P)$ and n for $|V|$.

For every edge in $[i, j] \in E$ we introduce two 0/1 variables x_{ij} and x_{ji}. If $x_{ij} = 1$ then the edge is directed from i to j. If $x_{ij} = 0$ then the edge is directed from j to i. So we have the constraint $x_{ij} + x_{ji} = 1$ for every $[i, j] \in E$. For every vertex i we further introduce 0/1 variables h_{ij}, $0 \le k \le d$. These variables will be interpreted as follows:
For the computed AOF we will have $h_{ij} = 1$ iff $deg(i) = j$ for this AOF. Hence the h-vector of P is computed by $h_j = \sum_{i=1}^{n} h_{ij}$, $0 \le k \le d$.

From the remarks following Lemma 1 we know that we have to find a cycle-free orientation of G which minimizes the quantity $\sum_{j=0}^{d} 2^j h_j$. The following ILP-model provides an AOF by the orientation defined via the variables x_{ij}:

$$\text{(H-AOF)} \qquad \text{Minimize} \qquad \sum_{j=0}^{d} 2^j h_j$$

subject to

$$\sum_{j=0}^{d} h_{ij} = 1 \qquad 1 \leq i \leq n \tag{1}$$

$$h_0 = 1 \tag{2}$$

$$\sum_{i=1}^{n} h_{ik} - h_k = 0 \qquad 0 \leq k \leq d \tag{3}$$

$$h_k - h_{d-k} = 0 \qquad 0 \leq k \leq \frac{d}{2} \tag{4}$$

$$h_k - h_{k-1} \geq 0 \qquad 1 \leq k \leq \frac{d}{2} \tag{5}$$

$$x_{ij} + x_{ji} = 1 \qquad \forall [i,j] \in E, i < j \tag{6}$$

$$\sum_{j=1}^{d} j h_{ij} + \sum_{(i,j) \in E} x_{ij} = d \qquad 1 \leq i \leq n \tag{7}$$

$$\sum_{(i,j) \in C} x_{ij} < k \qquad \begin{array}{l}\text{for every directed cycle } C \text{ with } k \\ \text{arcs whose underlying edges are in } E\end{array} \tag{8}$$

$$x_{ij} \in \{0,1\} \qquad \forall [i,j] \in E, i < j \tag{9}$$

The meaning of the constraints is as follows:

(1) Every vertex has exactly one value for the degree.
(2) h_0 is always 1.
(3) For every possible degree the number of vertices with this degree is determined.
(4) Dehn-Sommerville equation (Symmetry of the h-vector).
(5) Monotonicity of the first half of the h-vector.
(6) The orientation is well defined.
(7) Assigning correct vertex degree.
(8) The orientation of G is cycle-free.

It is clear that this model computes some AOF. The model is a Linear Ordering Problem with additional constraints. For actually solving it we solve the relaxed problem omitting constraint (8). The solution is tested for the existence of cycles and violated constraints of type (8) are added to the model. This is continued until no more cycle exists. We did not invest in using more of the sophisticated methods of [Rei85] to solve the problem.

In order to test the vertex set S of some quasifacet we have to extend the model by the following constraints:

$$x_{ij} = 1 \qquad \forall (i,j) \in E, i \in S, j \notin S \qquad (10)$$

If there exists a solution for this problem and the value of the objective function is equal to the value of the objective function of (H-AOF) then we have proved that S is a vertex set of a facet of P due to Theorem 2.

It should be noted that constraints (4) and (5) are redundant. However, it turned out that they substantially improve the performance of our code.

3. Computational experience

We have tested our algorithm RECONSTRUCT on several simple polytopes. The first two tables show our results. In all cases we were able to compute all facets. For all examples Conjecture 1 was correct. This means that in no case we had to use our ILP-model. We tested it anyway for making sure that it also works for non-trivial examples.

The first table deals with polytopes which were randomly generated. We generated the facet normals on the unit sphere and computed the combinatorial structure with the code PORTA [Chr91].

d	v	f	backtrack	CPU-time	decision	CPU-time all
3	26	15	0	0.06	1193	0.50
4	51	15	0	0.10	908	1.04
5	102	15	0	0.20	634	1.33
6	137	15	0	0.43	932	2.53
3	36	20	0	0.08	11775	5.87
4	82	20	14	0.21	6070	8.56
5	156	20	0	0.46	4115	11.04
3	46	25	4	0.12	105253	64.24
4	108	25	8	0.38	38152	67.66
7	160	15	0	0.50	9565	35.06
8	372	17	0	3.63	13305	105.81
8	585	17	0	8.90	18453	200.11
8	729	18	0	19.30	323598	4306.00

TABLE 1. Randomly generated simple polytopes

In Table 2 we consider the 4- and 5-cube and the following polytopes from the literature:

kleew: The only simple 4-polytope with 9 facets having graph-diameter 5 [KlW67].

d	v	f	backtrack	CPU-time	decision	CPU-time all	
4	16	8	0	0.05	24	0.06	4-cube
4	27	9	0	0.03	32	0.06	kleew
4	28	10	0	0.05	62	0.11	kleinsch
4	35	10	0	0.10	74	0.16	altsh
4	48	12	2	0.10	240	0.38	lock
5	32	10	0	0.06	90	0.14	5-cube

TABLE 2. Simple polytopes from the literature

kleinsch: The graph of the first known polytope whose
realization space is disconnected [BEK84].

altsh: The polar of a polytope whose polytopability had
been open for some time [Alt77].

lock: The polar of the first known polytope which was
not vertex-decomposable [Loc77].

In both tables the meaning of the column headers is the same. The columns d, v and f denote the dimension, number of vertices and number of facets of the simple polytope, respectively. If Conjecture 1 is valid we only have to find one quasifacet for each call of the algorithm AVALANCHE. This means that only some backtracking steps are needed to correct the decisions of decide for finding the only quasifacet. In column 4 and 5 the number of these backtrack steps and the corresponding computation time (CPU seconds on a SUN sparcstation 20 with 128 MB) is listed. If conjecture 1 is false we have to search the whole decision tree. The number of decisions made for finding all quasifacets is given in column 6. The overall computation time is shown in column 7. If conjecture 1 is correct the CPU time in column 5 solves the problem.

The quality of the procedure decide can be seen in the number of backtrack steps needed to compute only the first quasifacet for every call of the algorithm AVALANCHE. In most of the cases there was no backtrack step needed. This means that all decisions made by decide were correct.

It may be surprising that the 3-polytopes took many calls of decide. The reason is that there are many cycles in 3-polytopes which do not yield facets.

Tables 3 and 4 summarize the results for the ILP-model. The columns 1-3 denote the dimension, the number of vertices and the number of facets of the simple polytope, respectively. The h-vector is computed with the knowledge of $G(P)$ as described in our ILP-model. Column 5 shows the number of faces of the polytope. This is also the objective value of our ILP-model. The last column displays the computation time in CPU-seconds. As LP-solver we used CPLEX 3.0.

For some of the larger polytopes in Table 1 we stopped the ILP-model after an hour of computing time. Investing more knowledge from Combinatorial Optimization would certainly help to take care of them, too.

d	v	f	h-vector	faces	CPU-time
3	26	15	1 12 12 1	81	0.28
4	51	15	1 11 27 11 1	235	5.32
5	102	15	1 10 40 40 10 1	693	33.71
3	36	20	1 17 17 1	111	0.31
4	82	20	1 16 48 16 1	369	1.51
5	156	20	1 15 62 62 15 1	1047	656.51
3	46	25	1 22 22 1	141	0.35
4	108	25	1 21 64 21 1	483	2.64

TABLE 3. Computing the h-vector and an AOF for randomly generated polytopes

d	v	f	h-vector	faces	CPU-time	
4	16	8	1 4 6 4 1	81	0.21	4-cube
4	27	9	1 5 15 5 1	127	0.27	kleew
4	28	10	1 6 14 6 1	133	0.77	kleinsch
4	35	10	1 6 21 6 1	161	2.93	altsh
4	48	12	1 8 30 8 1	217	2.68	lock
5	32	10	1 5 10 10 5 1	243	0.37	5-cube

TABLE 4. Computing the h-vector and an AOF for polytopes from literature

References

[Alt77] A. Altshuler: *Neighbourly 4-polytopes and neighbourly combinatiorial 3 manifolds with ten vertices*, Canad. J. Math. 29 (1977), 400–420.

[BEK84] J. Bokowski, G. Ewald, P. Kleinschmidt: *On Combinatorial and affine automorphisms of polytopes*, Israel Journal of Mathematics, Vol. 47, Nos 2-3 (1984), 123–130.

[BlM87] R. Blind, P. Mani-Levitska: *On puzzles and polytope isomorphisms*, Aequationes Math. 34 (1987), 287–297.

[Chr91] T. Christof: *PORTA – A Polyhedron Representation Transformation Algorithm*, version 1.3, written by T. Christof (Uni Heidelberg), revised by A. Loebel and M. Stoer (ZIB Berlin); available from the ZIB electronic library ELIB via elib@zib-berlin.de

[Kal88] G. Kalai: *A simple way to tell a simple polytope from its graph*, J. Combinatorial Theory, Ser. A (1988), 381–383.

[KlW67] V. Klee, D. Walkup: *The d-step conjecture for polyhedra of dimension d < 6*, Acta Math., 133 (1967), 53–78.

[Loc77] E. R. Lockeberg: *Refinements in Boundary Complexes of Polytopes*, Ph.D. Thesis, University College London, (1977).

[Rei85] G. Reinelt: *The linear ordering problem: Algorithms and Applications*, Research and exposition in mathematics, Heldermann Verlag, Berlin (1985).

Hans Achatz and Peter Kleinschmidt
University of Passau
Department of Business Administration and Economics,
D-94030 Passau, Germany
[achatz,kleinsch]@winf.uni-passau.de

Reconstructing a Non-simple Polytope from its Graph

Michael Joswig

Abstract. A well-known theorem of Blind and Mani [4] says that every simple polytope is uniquely determined by its graph. In [11] Kalai gave a very short and elegant proof of this result using the concept of acyclic orientations. As it turns out, Kalai's proof can be suitably generalized without much effort. We apply our results to a special class of cubical polytopes.

1. Introduction

Polytopes arise as sets of admissible solutions for linear optimization problems. Most practical applications rely on Dantzig's Simplex Method to solve a given linear program. This algorithm starts at any vertex of the polytope and walks along improving edges to the optimum vertex. Thus the structure of the vertex-edge-graph is of considerable importance for the performance of the Simplex Method. There is reason to believe that geometric insight into the relationship between a polytope and its graph will help to solve some of the many unanswered questions about the worst-case running time of the Simplex Algorithm. For an overview over facts about graphs of polytopes see Ziegler [13, Chapter 3].

In this context it is natural to ask how far the graph of a polytope determines the (combinatorial) structure of the polytope. The known answer is: not very much, in general. For instance, there is a class of simplicial d-polytopes with n vertices, called *cyclic polytopes*, whose vertex-edge-graph is the complete graph on n vertices, which is the same as the graph of an $(n-1)$-simplex. In particular, in general, it is not even possible to determine the dimension of a polytope from its graph. See also Grünbaum [9, Chapter 12].

The situation is quite different for simple polytopes, as it has been shown by Blind and Mani [4]. Simple polytopes with isomorphic graphs are combinatorially equivalent. The purpose of this paper is to show how much information in addition to the graph is necessary in order to be able to reconstruct a general polytope. The first part of the paper closely follows Kalai's proof [11] of the theorem of Blind and Mani mentioned above. For questions concerning the complexity of the reconstruction procedure based on Kalai's proof see Achatz and Kleinschmidt [1].

The second part of this paper is devoted to a special class of cubical polytopes. It is shown that *capped* polytopes in the sense of Blind and Blind [3] can be reconstructed from their dual graphs.

Recently, Cordovil, Fukuda, and Guedes de Oliveira [6] proved a Blind-Mani type theorem in the context of oriented matroids. And we should also mention a result of Björner, Edelman, and Ziegler [2]: A zonotope can be reconstructed from its graph. The proofs in both papers rely on entirely different techniques than the ones used here.

2. The general reconstruction scheme

Let P be an arbitrary polytope and Γ_P its vertex-edge graph. For each vertex v of P, let η_v be the map from the set $\mathcal{F}(v)$ of facets containing v to the set of subsets of the set $\Gamma_P(v)$ of edges through v, such that each facet F is mapped onto the set of edges through v which are contained in F. We call the family of images $\{\eta_v(\mathcal{F}(v)) \mid v \text{ vertex of } P\}$ the *edge labeled vertex figures* of P. Observe that the vertex v can be recovered from the set $\eta_v(\mathcal{F}(v))$ of edge sets.

Lemma 2.1. *The face lattice of each vertex figure is determined by the edge labeled vertex figures.*

We follow Kalai's approach by examining acyclic orientations of Γ_P, see also Ziegler [13, Section 3.4]. A *good* acyclic orientation has precisely one sink on each non-empty face. An *abstract objective function* of P is a good acyclic orientation of Γ_P which also has precisely one source on each non-empty face. For simple polytopes it is known that the good acyclic orientations are precisely the *abstract objective functions* and also the *shellings* of the boundary of the dual (simplicial) polytope. This is a consequence of Kalai's proof. Note, however, that the graph of a non-simple polytope may have an acyclic orientation which is *not* an abstract objective function; see the example in Figure 1.

An induced subgraph Σ is called *initial* with respect to an acyclic orientation if there is no directed edge pointing to a vertex in Σ from any vertex of the complement of Σ. For an example of an initial subgraph with respect to a good acyclic orientation consider the following. Take an arbitrary linear objective function with minimal face M. Tilting the linear objective function slightly into general position induces a good acyclic orientation (and even an abstract objective function). The induced subgraph on M is initial. In particular, good acyclic orientations always exist.

In order to determine the combinatorial structure of P we have to find out which subsets of the vertex set form facets. Call a non-empty induced subgraph Φ of Γ_P an *F-subgraph* if it has the following properties:

(i) There is a good acyclic orientation O of Γ_P such that Φ is initial with respect to O.

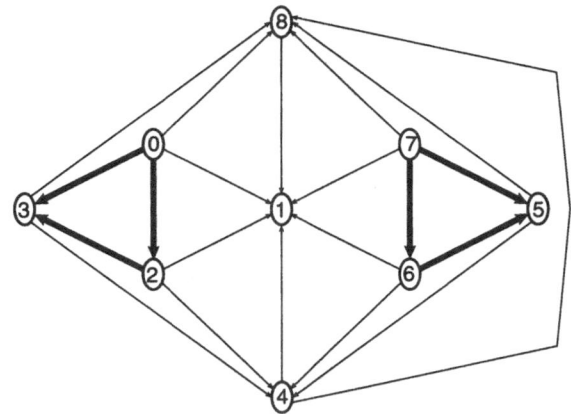

FIGURE 1. A good acyclic orientation of the graph of a simplicial 3-polytope which is not an abstract objective function. The
vertex 1 is the unique sink, while the vertices 0 and 7 are sources.
The subgraph corresponding to the facets $\{0, 2, 3\}$ and $\{5, 6, 7\}$,
respectively, is initial with respect to the acyclic orientation.

(ii) For each vertex v of Φ there is a (unique) facet $F \in \mathcal{F}(v)$ such that $\eta_v(F) = \Phi(v)$, where by $\Phi(v)$ we denote the set of edges through v in Φ.
(iii) The subgraph Φ is minimal with respect to inclusion among the induced subgraphs of Γ_P satisfying the properties above.

Note that the minimality condition enforces the connectedness of an F-subgraph. See again the example in Figure 1 for an induced subgraph of the graph of a 3-polytope satisfying (i) and (ii), but not (iii).

Lemma 2.2. *The F-subgraphs of Γ_P are precisely the subgraphs of facets of P.*

Proof. Clearly, the induced subgraph on the vertex set of any facet is an F-subgraph.
    For the converse let Φ be an F-subgraph. There is a good acyclic orientation O of Γ_P such that Φ is initial with respect to O. Choose a sink s among the vertices of Φ. By assumption there is a facet $F \in \mathcal{F}(s)$ with $\eta_s(F) = \Phi(s)$. Because O is good, the vertex s is the unique sink of F. Moreover, all the vertices of F are contained in Φ because Φ is initial with respect to O. Comparing Φ with the F-graph $\Gamma_F \leq \Gamma_P$ corresponding to F yields $\Phi = \Gamma_F$ due to the minimality of Φ.
□

    The proof of the preceding lemma shows that it is not necessary to require the uniqueness of the facet in (ii) of the definition of an F-subgraph.

Theorem 2.3. *A polytope can be reconstructed from its graph and its edge labeled vertex figures.*

Proof. Consider all possible acyclic orientations of the graph Γ_P. In view of Lemma 2.2 it suffices to exhibit all good acyclic orientations of Γ_P.

Fix an acyclic orientation O and a vertex v. We want to compute the number $f^O(v)$ of faces in which the vertex v is a sink with respect to O. This is the number of faces containing v built from edges incoming at v only. All faces through v can be enumerated because the face lattice of the vertex figure is known, cf. Lemma 2.1. Filter these faces for the incoming edge condition. Note that for P simple the number $f^O(v)$ solely depends on the in-degree of v with respect to O, which is not true for general polytopes. In particular, we do not have a formula in closed form.

Let $f^O = \sum_v f^O(v)$ and let f be the number of non-empty faces. We proceed as Kalai in his proof. As each non-empty face has at least one sink we have $f^O \geq f$. Further, O is good if and only if $f^O = f$. Because good acyclic orientations always exist we have that $f = \min\{ f^O \mid O \text{ acyclic orientation on } \Gamma_P\}$.   □

For P a simple d-polytope, the edge labeled the vertex figure of the vertex v coincides with the set $\binom{\Gamma_P(v)}{d-1}$, that is, the set of subsets of $\Gamma_P(v)$ of cardinality $d-1$. By iterated intersection they generate the boolean lattice $[2]^d$ (top element removed). Each vertex figure of a simple d-polytope is a $(d-1)$-simplex.

Corollary 2.4. *Every simple polytope can be reconstructed from its graph.*

3. Cubical polytopes

Simple polytopes are precisely the duals of simplicial polytopes. From a certain perspective cubical polytopes behave somewhat similar to simplicial polytopes. So the following conjecture is tempting; see also [6, Problem 3.3].

Conjecture 3.1. *Every cubical polytope can be reconstructed from its dual graph.*

The *dual graph* Δ_P of a polytope P is the vertex-edge graph of the dual polytope of P, i.e. $\Delta_P = \Gamma_{P^{\text{dual}}}$. The nodes and edges of Δ_P correspond to facets and ridges (codimension 2 faces) of P.

Note that, like for simplicial polytopes, there is no hope for cubical polytopes to be reconstructible from their graph: In [10] it is shown that there are cubical d-polytopes with the graph of the n-dimensional cube for arbitrary $n > d$. The *neighborly cubical polytopes* constructed in loc. cit. can be seen as cubical analogues of the (simplicial) cyclic polytopes mentioned in the introduction.

By definition every facet of a cubical d-polytope is a $(d-1)$-cube. In particular, every facet contains $2(d-1)$ ridges, i.e. the dual graph of a cubical d-polytope is $2(d-1)$-regular. A vertex figure of any vertex in a dual-to-cubical d-polytope is a $(d-1)$-dimensional cross polytope. In view of Theorem 2.3 in order to prove the conjecture it is sufficient to identify how the neighbors of any facet in the dual graph can be related to the vertices of a cross polytope. The main problem is that the induced subgraph among the neighbors in the dual graph is not the graph of

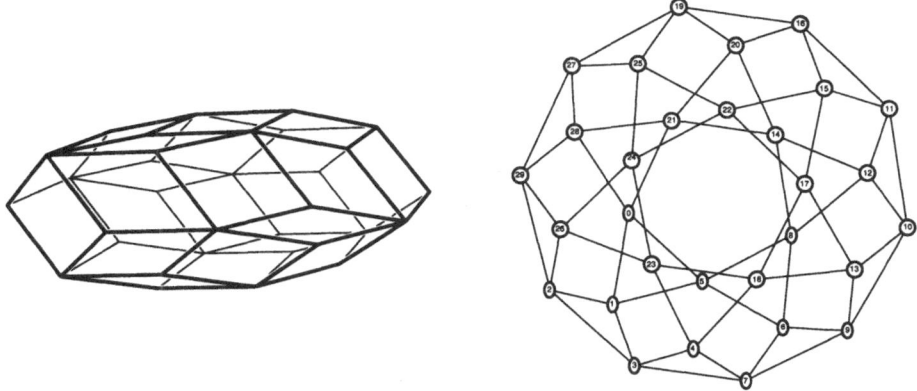

FIGURE 2. A 3-dimensional cubical zonotope (with 6 zones) and its dual graph.

a cross polytope, in general. Usually, there are many edges missing. For instance, in the example of a cubical 3-polytope in Figure 7 the induced subgraphs of the facets numbered 1, 7, 8, and 9, are totally disconnected.

We will show that Conjecture 3.1 holds for a very special class of cubical polytopes: A cubical polytope P is called *capped* over a cubical polytope Q if there is a combinatorial cube C such that $P = Q \cup C$ and $Q \cap C$ is a facet of Q. The unique facet of P which does not contain any vertex of Q is called the *cap* of P with respect to Q. A cubical polytope is called *capped* if it is obtained from iterated capping starting with a combinatorial cube; see Blind and Blind [3].

A few remarks on capped cubical polytopes. The property of being capped is *not* a combinatorial property as can be seen from Figure 3. For an example of a cubical polytope which is not combinatorially equivalent to a capped polytope, see the zonotope in Figure 2.

Before we will study dual graphs of cubical polytopes it is helpful to collect a few results about the *graphs* of cubical polytopes (and even more general graphs).

A pure polytopal complex is called *cubical* if all its facets are combinatorial cubes. A *cubical sphere* is a cubical complex which is homeomorphic to a sphere. We want to introduce the notion of *constructibility* of a pure cubical complex inductively. Every cube is *constructible*. If A and B are constructible d-dimensional cubical complexes, $A \cup B$ is a cubical complex, and $A \cap B$ is pure $(d-1)$-dimensional constructible, then $A \cup B$ is *constructible*. The *graph* of a complex is its 1-skeleton. Note that a constructible complex (and thus also its graph) are necessarily connected, as a straightforward inductive argument shows.

The following observation is due to Günter M. Ziegler.

Proposition 3.2. *The graph of a constructible cubical complex of dimension at least 2 does not contain any odd cycles. In particular, its graph is bipartite.*

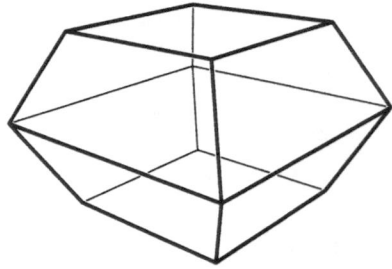 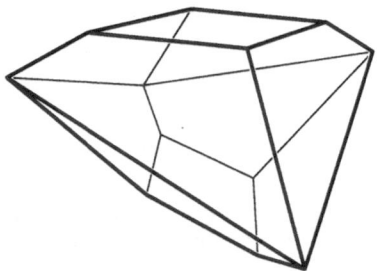

FIGURE 3. A capped and a not capped cubical polytope which are combinatorially equivalent. The polytope on the right is not capped because the quadrangle consisting of the 4 vertices around the waist is not planar.

Proof. No cube contains any odd cycle.

Assume there is a constructible cubical complex C which contains an odd cycle γ and which is not a cube. Then there are constructible cubical complexes A, B such that $A \cup B = C$. By an induction on the number of construction steps necessary to build up C we can assume that γ is contained in neither A nor B. That is, γ passes through the intersection $A \cap B$, which is connected, because of our assumption on the dimension. It is conceivable that γ enters and leaves $A \cap B$ several times. Pick two vertices x, y on γ in $A \cap B$ such that neither half of γ, obtained from cutting at x and y, is contained in $A \cap B$. If there is only one vertex in the intersection, then either A or B contains an odd cycle and we are done. Choose a path π in $A \cap B$ between x and y. Depending on the parity of the length of π, combining π with either half of γ yields an odd cycle γ'. Observe that γ' enters and leaves the intersection one times less than γ. An obvious induction now gives the result.   □

The boundary of a triangle is a 1-dimensional constructible cubical complex. This shows that the assumption concerning the dimension is necessary.

By a theorem of Bruggesser and Mani [5] the boundary of a polytope is known to be shellable and thus constructible. In particular, the boundary of a cubical polytope is a constructible cubical complex. So the above result implies that the graph of a cubical d-polytope, where $d \geq 3$, does not contain any triangle.

Two disjoint facets of a cube are called *opposite*.

Lemma 3.3. *Let A, B, C be facets in a cubical d-polytope P, where $d \geq 3$, such that $A \cap B$ and $B \cap C$ are opposite ridges in B.*

Then $A \cap C$ is not a ridge.

Proof. Assume $A \cap C$ is a ridge, cf. Figure 4.

Choose an edge e_A in A which is not contained in B or C. Let $(x, y) = e_A$ with $x \in C$ and $y \in B$. There is a unique edge e_B through y which is contained in B, but not contained in A. Now choose a linear objective function λ such that:

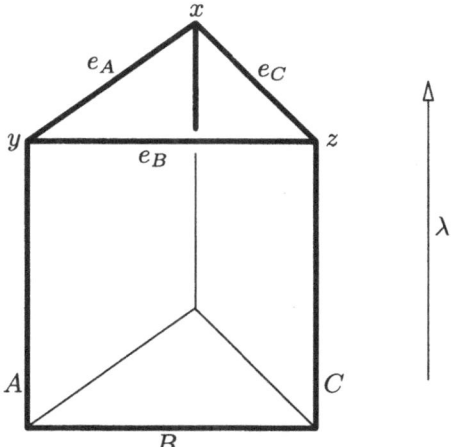

FIGURE 4. Three cubical facets and a linear function.

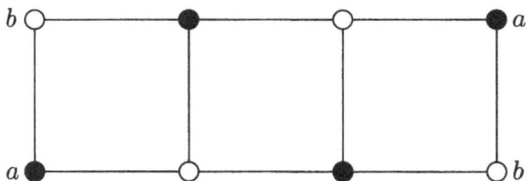

FIGURE 5. A cubical complex homeomorphic to the Möbius strip.

(i) the edge e_B is the unique maximal face of B with respect to λ, and
(ii) the function λ attains the same value on e_B and at x.

Let m_A be the maximal face of A with respect to λ. Clearly, m_A contains the edge e_A. If m_A were strictly greater than e_A, then the intersection $m_A \cap B$ would be strictly greater than y. This would contradict the maximality of e_B. We have $m_A = e_A$.

Denote by z the vertex of e_B which is contained in C. Let m_C be the maximal face of C with respect to λ. Now m_C contains the points x and z. As in the argument above, $m_C \supsetneq \{x, z\}$ contradicts the maximality of e_B in B. We conclude that $m_C = \{x, z\}$ is an edge. The vertices x, y, z form a triangle in the graph of the cubical polytope P. Due to Proposition 3.2 we arrive at the final contradiction. □

Note that there *are* (abstract) cubical complexes with a bipartite graph which do violate the conclusion of the preceding lemma: Take a Möbius strip built out of three quadrangles, cf. Figure 5.

Michael Joswig

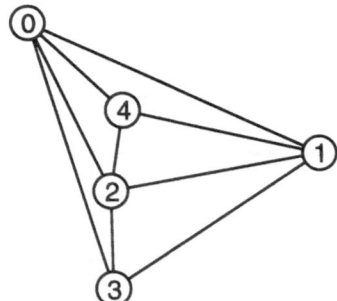

FIGURE 6. The graph of a 3-polytope with edge $(1,2)$ induced
among the neighbors of 0.

Dualizing Lemma 3.3 yields that, if e is an edge between two neighbors of
a vertex v in a dually cubical polytope, then e and v form a triangular face. For
general polytopes it may happen that the induced subgraph on the neighbors of
a given vertex v has edges which are not contained in any proper face through v,
see the example in Figure 6.

A further consequence of the above lemma is that the induced subgraph
among the neighbors in the dual graph of a cubical d-polytope is a subgraph
of a complete graph on $2(d-1)$ vertices minus a perfect matching. This means
that finding the edge labeling of the vertex figure of a facet in the dual graph is
particularly easy for a cubical polytope.

Lemma 3.4. *Let P be a cubical d-polytope, F one of its facets, Ω the set of neighbors of F in Δ_P. Then the set*

$$\left\{ \{N_1, \ldots, N_{d-1}\} \;\middle|\; N_i \in \Omega, \begin{array}{l} \text{any two facets } N_i \text{ and } N_j \\ \text{intersect non-trivially in } P \end{array} \right\}$$

is the edge labeling of the vertex figure of F in P^{dual}.

Differently phrased, Lemma 3.4 says that it suffices to recognize the antipodal
pairs among the neighbors of a facet in the dual graph. A similar result holds for
uniform oriented matroids, that is, oriented matroids which generalize cubical
zonotopes; cf. Cordovil, Fukuda, and Guedes de Oliveira [6, Theorem 3.1].

Lemma 3.5. *Let P be capped over the cubical polytope Q with cap F.
Then the induced subgraph Ω among the neighbors of F in Δ_P is a complete graph
minus a perfect matching. Moreover, Δ_Q is obtained from Δ_P by contracting $\Omega \cup
\{F\}$ to a single node. The edge labeled vertex figures of Q^{dual} determines the edge
labeled vertex figures of P^{dual} and vice versa.*

In Figure 7 the facet numbered 3 is a cap. The induced subgraph among its
neighbors is the graph of a quadrangle which is a 2-dimensional cross polytope.

If we already know that the cubical polytope is capped, then we also obtain
the converse of Lemma 3.5.

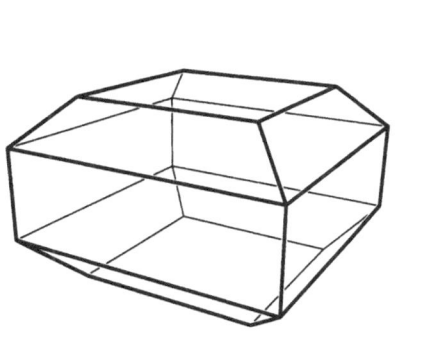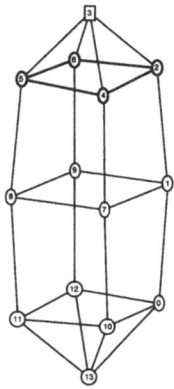

FIGURE 7. A capped cubical polytope and its dual graph.

Lemma 3.6. *Let P be a capped cubical d-polytope.*

If F is a facet with the property that the induced subgraph among the neighbors of F in Δ_P is a complete graph minus a perfect matching, then F is a cap.

Proof. Consider the set $\mathcal{R}$ of all the ridges contained in neighbors of F in Δ_P, but which are disjoint from F. By assumption the sublattice of the face lattice of P generated by $\mathcal{R}$ is isomorphic to the face lattice of a $(d-1)$-cube. We have to show that the ridges of F are the facets of a convex (combinatorial) cube, i.e. we have to show that the ridges in $\mathcal{R}$ are contained in a hyperplane.

A capped cubical d-polytope can also be seen as a d-dimensional (constructible) cubical complex. The facets of this complex are the (combinatorial) cubes which are attached one by one. Each facet of a capped cubical polytope is contained in a unique facet of the complex, i.e. a unique d-cube. Say, F is a facet of the d-cube C. Then the ridges in $\mathcal{R}$ are contained in the unique facet of C which is opposite to F. □

If we have a capped cubical polytope P, then Lemma 3.6 allows us to detect a cap by looking at the dual graph. Applying Lemma 3.5 then allows us to remove the cap, thus yielding a capped cubical polytope with fewer facets. Iterating this procedure we gather the edge labeled vertex figures of P. Now Theorem 2.3 gives the desired result.

Theorem 3.7. *Every capped cubical polytope can be reconstructed from its dual graph.*

Note that the property in Lemma 3.6 does not characterize a capping. If the induced subgraph among the neighbors of some facet F happens to be complete minus a perfect matching, then the ridges contained in the neighbors of F which are disjoint from F are combinatorially isomorphic to the boundary of a cube, but they are not necessarily contained in a hyperplane. An example for this situation can be seen in Figure 3 (right).

Acknowledgements. I am indebted to Manoj K. Chari, Volker Kaibel and Günter M. Ziegler for various helpful suggestions and to Friederike Körner for requiring a correction in Figure 1. The polytope pictures have been produced with Geomview [12] and Graphlet [7] via `polymake` [8].

References

[1] H. Achatz and P. Kleinschmidt, *Reconstructing a simple polytope from its graph*, this volume, 155–165.

[2] A. Björner, P.H. Edelman, and G.M. Ziegler, *Hyperplane arrangements with a lattice of regions*, Discrete Comput. Geom. **5** (1990), no. 3, 263–288.

[3] G. Blind and R. Blind, *Cubical 4-polytopes with few vertices*, Geom. Dedicata **66** (1997), no. 2, 223–231.

[4] R. Blind and P. Mani, *On puzzles and polytope isomorphisms*, Aequationes Math. **34** (1987), 287–297.

[5] H. Bruggesser and P. Mani, *Shellable decompositions of cells and spheres*, Math. Scandinav. **29** (1971), 197–205.

[6] R. Cordovil, K. Fukuda, and A. Guedes de Oliveira, *On the cocircuit-graph of an oriented matriod*, preprint, March 3, 1999.

[7] F.J. Brandenburg et al., *Graphlet, Version 5.0*, http://www.fmi.uni-passau.de/Graphlet/, Jan 11, 1999.

[8] E. Gawrilow and M. Joswig, *polymake: a framework for analyzing convex polytopes*, this volume, 43–74.

[9] B. Grünbaum, *Convex polytopes*, 1967.

[10] M. Joswig and G.M. Ziegler, *Neighborly cubical polytopes*, Discrete Comput. Geometry (to appear), math/9812033.

[11] G. Kalai, *A simple way to tell a simple polytope from its graph*, J. Combin. Th., Ser. A **49** (1988), 381–383.

[12] S. Levy, T. Münzner, and M. Phillips, *Geomview, Version 1.6.1*, http://www.geom.umn.edu/software/geomview/, Oct 30, 1997.

[13] G.M. Ziegler, *Lectures on polytopes*, Springer, 1998, 2nd ed.

Michael Joswig
Fachbereich Mathematik, MA 7–1
Technische Universität Berlin
Straße des 17. Juni 136, D-10623 Berlin, Germany
joswig@math.tu-berlin.de

A Revised Implementation of the Reverse Search Vertex Enumeration Algorithm

David Avis

Abstract. This paper describes an improved implementation of the reverse search vertex enumeration/convex hull algorithm for d-dimensional convex polyhedra. The implementation uses a lexicographic ratio test to resolve degeneracy, works on bounded or unbounded polyhedra and uses exact arithmetic with all integer pivoting. It can also be used to compute the volume of the convex hull of a set of points. For a polyhedron with m inequalities in d variables and known extreme point, it finds all bases in time $O(md^2)$ per basis. This implementation can handle problems of quite large size, especially for simple polyhedra (where each basis corresponds to a vertex and the complexity reduces to $O(md)$ per vertex). Computational experience is included in the paper, including a comparison with an earlier implementation.

1. Introduction

This paper describes *lrs* [1], a revised version of the reverse search vertex enumeration algorithm proposed by Fukuda and the author [2]. This implementation is a major improvement on *rs*, a program released by the author in 1992, and revised in 1994 [4], which was based on the original method described in [2]. The improvements in speed are attributable to many factors, including an improved pivot selection procedure to resolve degeneracy lexicographically, faster arithmetic based on integer rather than rational arithmetic, and optional cacheing to reduce backtrack pivots. The new implementation is more general, as it handles bounded and unbounded polyhedra, does not require non-negativity of variables, and performs automatic transformations to allow solution of facet enumeration and Voronoi diagram problems.

The main function of *lrs* is to find the vertices and extreme rays of a polyhedron described by a system of linear inequalities. A description of the theory underlying the current implementation of this function is the purpose of this paper. Additional functions of *lrs* are built on top, and are described briefly here. These include: facet enumeration, computation of Voronoi vertices, volume computation, estimation of the output size, and restart capability. The remainder of this introduction contains an informal description of how *lrs* works. Section 2 of the paper gives some background material and basic definitions. Section 3 describes dictionaries, their relationship to vertices and extreme rays and pivoting. Section 4 describes lexicographic pivoting and proves that each vertex and extreme ray is representable by a lex-positive basis. Section 5 is concerned with the unique representation of vertices and extreme rays by lex-min bases. Section 6 contains a description of reverse search, and an efficient implementation of the reverse function as used in *lrs*. Section 7 describes some implementation issues, including the integer pivoting formulae used. Section 8 describes briefly how the additional functions mentioned above are implemented. Section 9 concludes with some limited computational experience comparing the current version *lrs* to the earlier version *rs*. More extensive experience, and an empirical analysis of the various speedups is contained in Avis [7].

Briefly and informally, the reverse search algorithm works as follows. Suppose we have a system of m linear inequalities defining a d-dimensional polyhedron in $\mathbb{R}^d$ and a vertex of that polyhedron given by the indices of d inequalities whose bounding hyperplanes intersect at the vertex. These indices define a *cobasis* for the vertex. The complementary set of $m - d$ indices are called a *basis*. For any given linear objective function, the simplex method generates a path between *adjacent* bases (or equivalently cobases) which are those differing in one index. The path is terminated when a basis of a vertex maximizing this objective function is found. The path is found by pivoting, which involves interchanging one of the hyperplanes defining the current cobasis with one in the basis. The path chosen from the initial given basis depends on the pivot rule used, which must be finite to avoid cycling. The original implementation *rs* used Bland's least subscript rule [8]. If we look at the set of all such paths from all bases of the polyhedron, we get a spanning forest of the graph of adjacent bases of the polyhedron. The root of each subtree of the forest is a basis of an optimum vertex. The reverse search algorithm starts at each root and traces out its subtree in depth first order by *reversing* the pivot rule.

The algorithm is particularly easy if the polyhedron is *simple* (non-degenerate): each vertex lies on exactly d hyperplanes, and so has a unique basis. The spanning forest has one component, which is a spanning tree of the skeleton of the polyhedron, and each vertex is produced once. An example of such a polyhedron is the cube, and Figure 1 shows a possible reverse search tree for it. A complication of the original implementation was the need to handle a degenerate starting vertex. A dual form of Bland's rule had to be implemented in order to compute all bases of this vertex. Each of these bases was then used as the start for a (primal) reverse search. A serious drawback of the algorithm as originally implemented was that it computed all bases of the input polyhedron. For many polyhedra encountered, especially combinatorial polytopes, the number of bases is much larger than the number of vertices.

The standard approach to reducing the number of bases is perturbation: make small changes to the input data so that the resulting polyhedron is simple. The resulting perturbed polyhedron will typically have more vertices, but far fewer bases, than the original polyhedron. Numerical perturbation was implemented for

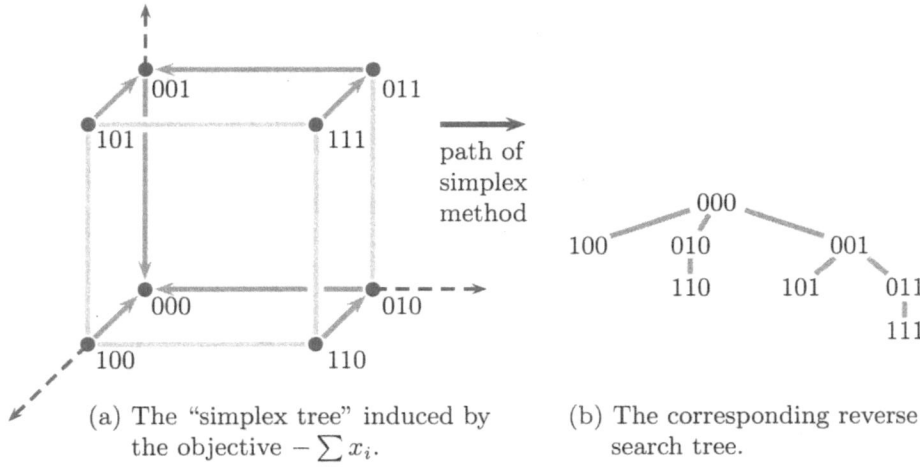

(a) The "simplex tree" induced by the objective $-\sum x_i$.

(b) The corresponding reverse search tree.

FIGURE 1

rs, but abandonded because: (i) perturbed vertices have to be transformed back to give true vertices; (ii) the answer may no longer be correct as vertices may be lost; (iii) perturbation increases the cost of the extended precision arithmetic; (iv) a vertex is output more than once.

The current version lrs resolves degeneracy by use of the well-known lexicographic pivot selection rule for the simplex method. This rule is defined for a subset of the bases, known as lex-positive. The subgraph of lex-positive bases forms a connected subgraph of the basis graph which covers all vertices of the polyhedron. Furthermore an objective function can be chosen so that the simplex method initiated at any lex-positive basis terminates at a unique lex-positive optimum basis. If we initiate the reverse search method at this basis and reverse the lexicographic pivot rule we generate a spanning tree of the graph of all lex-positive bases. This is the core of lrs.

As we will see in Section 5, the lex-min basis for each vertex is lex-positive. We can test the property of being a lex-min basis quickly, and report only lex-min bases. Lexicographic pivoting is sometimes referred to as symbolic perturbation, since the graph of lex-positive bases can be viewed as the skeleton of a simple polyhedron obtained by perturbation of the original polyhedron. Indeed, it is often described by adding powers of a small indeterminate ϵ to the right hand side of each inequality to resolve degeneracy. However it can be implemented without any change to the original data, and so all of the objections (i)–(iv) above are resolved. Furthermore we can use the same method for the dual problem of computing the facets of the convex hull of as set of vertices and extreme rays. In this dual context, properly interpreted, the graph of lex-positive bases corresponds to a triangulation of the convex hull. Since the determinant of each of these simplices is known (see Section 7), an additional advantage is that we can obtain the volume of a polytope given by a set of vertices for the same cost as computing the facets of its convex hull.

A remarkable feature is that no additional storage is needed at intermediate nodes in the tree. Going deeper in the tree we explore all valid "reverse" pivots in order by basic index from any given intermediate node. For backtracking, we can use the pivot rule to return us to the parent node along with the current pivot indices. From there it is simple to continue by considering the next basic index as a "reverse" pivot, etc. The algorithm is therefore non-recursive and requires no stack or other data structure. The output is produced as a duplicate free stream (see Section 5), which may be useful even if the computation is too large to complete. Empirically, about 80–90% of the running time of *lrs* is spent pivoting. Since backtracking generates exactly half of the pivots, considerable savings can however be obtained by "cacheing" dictionaries along the current path from the root.

Although *lrs* is a large improvement on *rs*, it is far from an efficient general solution to the vertex enumeration problem. Such a solution should reasonably be required to generate all vertices in time polynomial in the input and output size. Currently no such algorithm is known to exist. Examples contained in Avis, Bremner and Seidel [6] show that all pivot algorithms using numeric or symbolic perturbation may behave extremely badly: the number of bases computed can be super-polynomial in the number of vertices. This is born out in practice for combinatorial polytopes. *lrs* is efficient for vertex enumeration of simple (or near-simple) polyhedra, or dually for facet enumeration of simplicial (or near-simplicial) polyhedra. Recently Bremner, Fukuda and Marzetta [9] developed an ingenious primal-dual method for vertex enumeration of highly degenerate simplicial polytopes that satisfy a hereditary property. It works by simulating the reverse search tree generated by *lrs* for the (easy) dual facet enumeration problem for simplicial polytopes. Dually, this method can be used for facet enumeration of simple polytopes. For polytopes with zero-one vertices, a polynomial time vertex enumeration algorithm was recently announced by Bussieck and Luebbecke [12].

lrs can be efficiently parallelized, as has been done by Brüngger, Marzetta, Fukuda and Nievergelt [10]. This parallel version has been used to solve some extremely large problems which do not seem solvable by other methods.

2. Background

For a general introduction to convex polyhedra, the reader is referred to Ziegler [20]. Throughout the paper we will assume that P is a d-dimensional polyhedron in $\mathbb{R}^d$. A classic result is that P can be represented in two ways. An $\mathcal{H}$-*representation* is given by an $(m \times d)$-matrix $\overline{A} = (\overline{a}_{i,j})$ and an m-vector $\overline{b} = (\overline{b}_i)$:

$$P = \{y \in \mathbb{R}^d \mid \overline{b} + \overline{A}y \geq 0\} \tag{2.1}$$

If $\overline{A}$ is minimal, that is no row can be deleted without changing P, then P has m facets, each defined by one of the inequalities in (2.1). A *vertex* $y \in \mathbb{R}^d$ is a point of P that satisfies an affinely independent set of d inequalities as equations. We assume throughout that P has at least one vertex, which implies that $m \geq d + 1$. An *extreme ray* $z \in \mathbb{R}^d$ is a direction such that for some vertex y and any positive scalar t, $y + tz$ is in P and satisfies some set of $d-1$ affinely independent inequalities as equations. Note that an extreme ray is unique only up to a positive scalar, since if z is an extreme ray then so is tz for any positive scalar t. An equivalent $\mathcal{V}$-*representation* of P is given by a minimal set of s vertices $y_1, \ldots, y_s$ and u extreme rays $z_1, \ldots, z_u$:

$$P = \left\{ y \in \mathbb{R}^d \mid y = \sum_{i=1}^{s} \lambda_i y_i + \sum_{j=1}^{u} \mu_j z_j, \ \lambda_i \geq 0, \ \mu_j \geq 0, \ \sum_{i=1}^{s} \lambda_i = 1 \right\}. \quad (2.2)$$

The *vertex enumeration problem* is to produce a $\mathcal{V}$-representation from an $\mathcal{H}$-representation, and the *facet enumeration problem* is to provide the reverse transformation. It is well known that these problems are essentially equivalent, see Section 8.1. In this paper we treat the problem primarily from the vertex enumeration perspective.

As an example, consider the unbounded 3-dimensional polyhedron P defined by the system of 8 inequalities:

$$1 + x_1 \geq 0$$
$$1 + x_2 \geq 0$$
$$1 - x_1 \geq 0$$
$$1 - x_2 \geq 0$$
$$1 - x_1 + x_3 \geq 0$$
$$1 - x_2 + x_3 \geq 0$$
$$1 + x_1 + x_3 \geq 0$$
$$1 + x_2 + x_3 \geq 0$$

It has a $\mathcal{V}$-representation given by the 5 vertices

$$(1, 1, 0), \ (-1, 1, 0), \ (1, -1, 0), \ (-1, -1, 0), \ (0, 0, -1)$$

and one extreme ray $(0, 0, 1)$. Note that the vertices do not have unique cobases. For example the vertex $(-1, -1, 0)$ can be defined by choosing any three of the first two and last two inequalities and replacing them by equations. The extreme ray has four representations, one with each of the first four vertices. For example $(1, -1, 0) + (0, 0, t)$ satisfies both the second and third inequalities as equations, and is in P for all $t \geq 0$. These representations with distinct vertices are called *geometric rays*.

3. Dictionaries

Much of the material in this section is adapted to the vertex enumeration setting from standard results in linear programming, see for example Chvátal [13] and Ignizio and Cavalier [18]. As with the simplex method, the essential calculations are performed on a *dictionary* derived from (2.1). We distinguish *decision* variables $x_1, \ldots, x_d$ and *slack* variables $x_{d+1}, \ldots, x_{m+d}$. It is easy to see that solutions to (2.1) can be put in 1-1 correspondence to solutions of

$$x_{d+i} = \bar{b}_i + \sum_{j=1}^d \bar{a}_{i,j} x_j, \qquad i = 1, \ldots, m,$$

$$x_{d+i} \geq 0, \qquad\qquad\qquad i = 1, \ldots, m,$$

by identifying $x_j = y_j$ for $j = 1, \ldots, d$. lrs is initiated from a vertex of P, which is either supplied by the user or is computed by solving a linear program over (2.1). It is convenient to order the rows of (2.1) so that the final d rows define this initial vertex. Since these rows are affinely independent, we can rewrite the above system of m equations as the equivalent system

$$x_i = b_i + \sum_{j=1}^d a'_{i,j} x_{m+j}, \qquad i = 1, \ldots, m. \tag{3.1}$$

for a suitable coefficients b_i and $a'_{i,j}$. For reasons that become clear later, we augment this system by adding the additional equation

$$x_0 = b_0 + \sum_{j=1}^d a'_{0,j} x_{m+j} \tag{3.2}$$

where $b_0 = 0$ and $a'_{0,j} = -1$, for $j = 1, \ldots, d$. From the vector $b = (b_0, \ldots, b_m)$ and $(m+1) \times d$ matrix $A' = (a'_{i,j})$ we form the $(m+1) \times (d+m+1)$ dimensional matrix

$$A = (I \quad -A')$$

where I is an $(m+1) \times (m+1)$ identity matrix. Then the augmented system of equations can be rewritten as

$$Ax = b. \tag{3.3}$$

For any matrix such as A, A_j refers to the j-th column of A, and A_J refers to the submatrix of columns of A indexed by J. We use similar notation for vectors. The notation A^i and A^I refers to row i and the submatrix of A with rows indexed by I respectively. Let B be an ordered $(m+1)$-*tuple* indexing a set of $m+1$ affinely independent columns of A, and let N be an ordered d-*tuple* indexing the remaining d columns. Then we may rewrite (3.3) as $A_B x_B + A_N x_N = b$, which is equivalent to the system

$$I \, x_B + A_B^{-1} A_N x_N = A_B^{-1} b \tag{3.4}$$

The system (3.4) is called a *dictionary*, and consists of $m+1$ rows.

The t-th row corresponds to the t-th index in B. All essential calculations involve manipulating this dictionary. The index set B is called a *basis*, and the index set N is called a *cobasis*. From any dictionary we obtain a *basic solution* x by setting $x_B = A_B^{-1}b$ and $x_N = 0$. B is a *feasible basis*, x is a *basic feasible solution*, and (3.4) is a *feasible dictionary* if:

(i) $\{0, \dots, d\} \subset B$, and
(ii) $x_i \geq 0$ for $i \in B$, $i = d+1, \dots, d+m$.

In other words, B contains x_0 and all decision variables, and x is non-negative for all slack variables. The next two propositions show how the vertices and extreme rays of P can be recovered from dictionaries.

Proposition 3.1. *Every vertex y of P can be extended to a basic feasible solution of (3.4), and every basic feasible solution x of (3.4) can be restricted to a vertex of P.*

Proof. Let y be a vertex of P. It is defined by choosing a a set of d inequalities in (2.1) and replacing them by equations. Let N be the set of all indices $i+d$ such that inequality i is chosen, and let $B = \{0, \dots, d+m\} - N$. Let $x_j = y_j, j = 1, \dots, d$, define $x_{d+i}, i = 1, \dots, m$ by (3.1) and x_0 by (3.2). Clearly $x_N = 0$ and (i) and (ii) are satisfied since y satisfies (2.1). Conversely, if x is a basic feasible solution, we define y by $y_j = x_j, j = 1, \dots, d$. Since $x_N = 0$, N indicates a set of d inequalities from (2.1) that are solved as equations to define y. Condition (ii) indicates y is feasible for (2.1), hence it is a vertex. $\square$

Proposition 3.2. *For every extreme ray z of P, there is a feasible basis B and index $s \in N$ such that letting $a = A_B^{-1}A_s$:*

(a) $z_i = -ta_i, 1 \leq i \leq d$ *and some* $t > 0$
(b) $a_i \leq 0, d+1 \leq i \leq m$.

Proof. From the definition, each extreme ray of P is a feasible direction $z \in \mathbb{R}^d$, so that for some vertex y and any positive scalar t, $y + tz$ satisfies a set of $d - 1$ inequalities from (2.1) as equations. Let N' index this set. Since P does not admit lines, this ray is terminated by one additional inequality being satisfied as an equation. Let s index this inequality and set $N = N' + s$. We obtain a set of d cobasic indices N which determine the vertex y of P. Let $a = A_B^{-1}A_s$. By (b) we can obtain feasible solutions $x = A_B^{-1}b - at$ by setting $x_{N'} = 0$ and $x_s = t$ for any positive scalar t. All these solutions satisfy inequalities indexed by N' as equations, so restricting to coordinates $1, \dots, d$ of $-a$, we obtain the extreme ray z. $\square$

Under the conditions of Proposition 3.2, we say that B *represents* the geometric ray $y + tz$. *lrs* computes feasible dictionaries from which vertices and extreme rays of P may be obtained. The basic operation is a *pivot* between two bases B and $\overline{B}$, which is defined by indices $r \in B$ and $s \in N$ by setting $\overline{B} = B - r + s$. This notation means that s replaces r in its position (say t) in the basis B. The operation is the computation of (3.4) for $\overline{B}$, which is done as follows. Let $a = A_B^{-1}A_s$.

Pivot operation

(a) Divide row t of (3.4) by a_t.
(b) For $i = 0, \ldots, t-1, t+1, \ldots, m$, subtract a_i times the new row t from row i of (3.4).

We discuss the mechanics in Section 7. The pivot is *feasible* if both x_B and $x_{\bar{B}}$ are basic feasible solutions, which implies that both x_r and x_s are slack variables. The decision variables do not play any active role in pivot selection, they are "carried" along merely to enable the vertex and extreme ray coordinates to be computed (they are columns of this part of the dictionary).

Let $B^* = \{0, \ldots, m\}$ and $N^* = \{m+1, \ldots, m+d\}$. This is the initial basis and cobasis, and the initial dictionary is given by (3.2) and (3.1), which we write:

$$I\, x_{B^*} + A_{N^*} x_{N^*} = b. \tag{3.5}$$

The first equation in this system is

$$x_0 + \sum_{i \in N^*} x_i = 0 \tag{3.6}$$

If we interpret x_0 to be the value of an objective function, then the basic solution of this dictionary with $x_{N^*} = 0$ has objective value zero. Since for all basic feasible solutions we have $x_j \geq 0$, $j \in N^*$, it is clear that the maximum of x_0 over all basic feasible solutions is zero, and the initial dictionary achieves this maximum. This initial dictionary is the root of the reverse search tree constructed by *lrs*. All other bases are obtained by pivoting from this dictionary. The initial basis inverse is the $(m+1) \times (m+1)$ identity matrix.

To perform lexicographic pivots from any feasible basis B, it will be necessary to have access to the basis inverse A_B^{-1}. This can be readily obtained from the dictionary (3.4) with basis B as follows. Let e^i be the unit $(m+1)$-vector indexed $0, \ldots, m$ with a one at index i.

Proposition 3.3. *Let B be a feasible basis. Then, for $i = 0, \ldots, m$, column i of A_B^{-1} is e^r if $i = B_r$ for some r, otherwise it is the column $A_B^{-1} A_i$.*

Proof. The dictionary (3.4) corresponding to B is obtained by standard row and column operations from the equivalent system (3.5). It follows from elementary linear algebra that the columns indexed by $B^* = \{0, \ldots, m\}$ in (3.4) are the basis inverse A_B^{-1}.   □

4. Lex-positive bases and lexicographic pivot selection

The earlier program *rs* computed all basic feasible solutions, and so by Proposition 3.1 this guarantees all vertices will be found. For degenerate polyhedra there are many more feasible bases than vertices, so it useful to identify a smaller set of bases that still cover all the vertices. We call a vector *lex-positive* if its first non-zero coordinate is positive. We call a feasible basis B *lex-positive* if each of the rows indexed $d+1,\ldots,m$ of the $(m+1) \times (m+1)$ matrix

$$D = (A_B^{-1}b \ A_B^{-1}) = \begin{pmatrix} \beta_0 & \alpha_{0,1} & \cdots & \alpha_{0,m} \\ & \cdot & & \cdot \\ & \cdot & & \cdot \\ & \cdot & & \cdot \\ \beta_m & \alpha_{m,1} & \cdots & \alpha_{m,m} \end{pmatrix} \tag{4.1}$$

is lex-positive. The initial basis B^* is lex-positive, since in this case (4.1) is $(\,b\ I\,)$, and the condition follows from the feasibility of B^*. Every feasible basis B contains indices $0,\ldots,d$ and so the columns with these indices in (4.1) remain unchanged, they are the first $d+1$ columns of the identity matrix. We will show that each vertex has a lex-positive basis. To do this we introduce lexicographic pivoting, which preserves the property that the basis is lex-positive.

Let B be a lex-positive basis. Let $s \in N$ and consider the column $a = A_B^{-1}A_s$. If $a_i \leq 0$, $d+1 \leq i \leq m$, this column defines an extreme ray, as described in Proposition 3.2. Otherwise, let t be the index such that D^t/a_t is the lexicographically minimum vector of

$$\left\{ \frac{D^i}{a^i} \ : \ a_i > 0, \ d+1 \leq i \leq m \right\} \tag{4.2}$$

Such a minimum is unique, because D has full row rank. Let r be the basic index in B corresponding to row t. We abbreviate this in the notation:

$$r = lexminratio(B,s)$$

In case (4.2) is empty and we set $r = 0$.

Proposition 4.1. *Given a lex-positive basis B and $s \in N$, let $r = lexminratio(B,s) \neq 0$.*

(a) $\overline{B} = B - r + s$ *is a lex-positive basis.*
(b) $s = lexminratio(\overline{B}, r)$.

Proof. Let t be the row of D corresponding to basic index $r \in B$, and set $a = A_B^{-1}A_s$. The pivot operation produces the matrix $\overline{D}$ for $\overline{B}$ from D according to the formulae:

$$\overline{D}^t = \frac{D^t}{a_t}, \qquad \overline{D}^i = D^i - a_i\overline{D}^t, \qquad 0 \leq i \leq m, \ i \neq t. \tag{4.3}$$

To prove the lex-positivity of $\overline{B}$ we need only consider rows indexed $d+1 \leq i \leq m$. Since each row of D is lex-positive, and $a_t > 0$, $\overline{D}^t$ is lex-positive. Therefore, for each i with $a_i \leq 0$, $\overline{D}^i$ is the sum of a lex-positive vector with either zero or a lex-positive vector, so it is lex-positive. For each i with $a_i > 0$ we can rewrite its equation above as

$$\frac{\overline{D}^i}{a_i} = \frac{D^i}{a_i} - \frac{D^t}{a_t}.$$

Since t was chosen as the lexicographically minimum vector in (4.2), it follows that the right hand side and hence $\overline{D}_i$ is lex-positive. This proves part (a). For part (b) set

$$\overline{a} = A_B^{-1} A_r.$$

In the dictionary with basis B, column r is the identity column e^t. Therefore it follows from the pivot operation that

$$\overline{a}_t = \frac{1}{a_t}, \quad \text{and} \quad \overline{a}_i = -a_i \overline{a}_t \text{ for } d+1 \le i \le m, \ i \ne t.$$

Using these definitions, for any i, $d+1 \le i \le m$, $i \ne t$ such that $\overline{a}_i > 0$ we can deduce from (4.3) that

$$\frac{\overline{D}^i}{\overline{a}_i} = \frac{D^i}{\overline{a}_i} + \frac{\overline{D}^t}{\overline{a}_t}.$$

Since both terms on the right hand side are lex-positive, we conclude that $\overline{D}^i/\overline{a}_i$ is lexicographically greater than $\overline{D}^t/\overline{a}_t$. Therefore row t is the minimizer for basis $\overline{B}$. This proves part (b), since $s \in \overline{B}$ is the index corresponding to row t of the updated dictionary. □

The lexicographic pivot rule for the simplex method for linear programming chooses pivots in the following way. Assuming the problem is to maximize x_0, the index $s \in N$ of the entering variable is chosen so that it has a negative coefficient in row zero of (3.4). If no such index exists, the current basic feasible solution is optimum. We then choose $r = lexminratio(B, s)$ as the index of the leaving variable. If no such index exists, the problem is unbounded. The proof of Proposition 4.1 shows that each pivot adds a positive multiple of the lex-positive row r to the objective function. Since the objective row increases lexicographically with each pivot, this proves no basis can repeat and the optimum (or an unbounded solution) must eventually be reached. Furthermore, if pivots are chosen in this way, an optimum solution will be found with a lex-positive basis, even though in general it may have many other bases that are not lex-positive. We may now strengthen Proposition 3.1.

Proposition 4.2. *Every vertex y of P can be extended to a basic feasible solution of (3.4) with a lex-positive basis.*

Proof. It is well known that for every vertex y of P there is a linear function which obtains its maximum over P at the unique point y. We use this function instead of the the the one specified in (3.2), and initiate the simplex method using lexicographic pivoting on the resulting dictionary (3.4) with the initial lex-positive basis B^*. By the above discussion, we will obtain the optimum solution y with a lex-positive basis. □

We call a feasible basis B and its corresponding dictionary *optimum* if row zero of $A_B^{-1} A$ is non-negative. As described above, this is the normal stopping criterion for the simplex method. An important feature of lex-positive bases is that there is a unique optimum dictionary, eliminating the need to initiate reverse search on multiple dictionaries. A similar result was obtained by Bremner et al. [9] using a different argument.

Proposition 4.3. B^* *is the unique optimum lex-positive basis.*

Proof. Row zero of the initial dictionary is the $m+d+1$ vector $(1, 0, \ldots, 0, 1, \ldots, 1)$ with m zeroes, derived from equation (3.6), where the cobasis is $N^* = \{m + 1, \ldots, m + d\}$. Suppose there exists another lex-positive optimum basis $B \neq B^*$. Let $k_1, \ldots, k_q$ be indices in $B - B^*$. Each index is necessarily in N^*. Consider row zero of the system (3.4). It is obtained from (3.6) by subtracting equations with variables indexed by $k_1, \ldots, k_q$ from (3.4), since these variables are eliminated from the cobasis N^*. The coefficients of variables $x_0, \ldots, x_m$ in these equations are the corresponding rows in (4.1), each of which is lex-positive by assumption. Since both B and B^* are optimum, $\beta_{k_j} = 0$. It follows that $(\alpha_{0,1}, \ldots, \alpha_{0,m})$ is lexicographically negative, so $\alpha_{0,j} < 0$ for some index $d + 1 \leq j \leq m$ (in fact, the first non-zero coefficient). This contradicts the optimality of B. $\square$

The results of this section are summarized in the following theorem.

Theorem 4.4. *For each vertex y of P there is a lex-positive basis B with basic feasible solution x such that $x_i = y_i$ for $i = 1, \ldots, d$. The simplex method initiated on the corresponding dictionary (3.4) generates a sequence of lex-positive bases terminating in the basis B^*.*

5. Lex-minimum bases of vertices and extreme rays

Even with lex-positive pivoting, for a non-simple polyhedron the same vertex will often be generated with many bases. Fortunately, a unique representative basis can easily be identified for each vertex. For a given basic feasible solution x of (3.4), its lexicographically smallest basis B is called *lex-min*.

Proposition 5.1. *B is lex-min for some basic feasible solution x if and only if there does not exist $r \in B$ and $s \in N$ such that*

(i) $r > s$,
(ii) $x_r = 0$, *and*
(iii) $(A_B^{-1} A_s)_r \neq 0$.

Proof. Clearly if B is lex-min we cannot satisfy (i)–(iii), for otherwise, the basis $\overline{B} = B - r + s$ has the same basic feasible solution x and is lexicographically smaller than B. On the other hand, suppose B is not lex-min. Let B_{min} be the lex-min basis with the same basic feasible solution x. Note that this implies that if $x_i \neq 0$, i is contained in both B and B_{min}. Let s be the smallest index in $B_{min} - B$. Since B is a basis there is some index $r \in B - B_{min}$ such that $\overline{B} = B - r + s$ is a feasible basis. By the choice of r and s we must have both (i) and (ii). Finally (iii) follows from the fact that r and s define a valid pivot for the dictionary with basis B. $\square$

Given a dictionary for B, conditions (i)–(iii) of Proposition 5.1 can be checked in $O(md)$ time.

Proposition 5.2. *The lex-min basis B of each basic feasible solution x is lex-positive.*

Proof. We must show the rows indexed $d+1, \ldots, m$ of (4.1) are lex-positive. Since x is feasible, we need only check rows with $\beta_i = 0$. Let $r \in B$ be a basic index corresponding to such a row. Suppose the smallest index j for which $\alpha_{i,j} \neq 0$ is such that $\alpha_{i,j} < 0$. Now if $r \leq m$ from Proposition 3.3 we have $\alpha_{i,r} = 1$, so $j < r$. On the other hand, if $r \geq m+1$, then trivially $j \leq m < r$. Also $j \in N$ since it indexes a column of (4.1) that is not a unit vector. Therefore $\overline{B} = B - r + j$ is a feasible degenerate pivot yielding a lexicographically smaller basis for the same feasible solution x. Since this is a contradiction, it must be that $\alpha_{i,j} > 0$ and so row i is lex-positive. $\qquad\square$

Note that the proof of the proposition implies that in a lex-min basis B, if $i \in B$ and $i \geq m+1$, then necessarily $x_i > 0$.

Next we consider extreme rays. Here the situation is more complicated, due to the existence of parallel rays incident to distinct vertices, known as geometric rays, as illustrated in Section 2. In a non-simple polyhedron, the same geometric ray may appear in many dictionaries, that is, it may have many cobasis representations with the same basic solution. It is not necessarily true that a geometric ray incident with a vertex y will appear in the lex-min dictionary for this vertex. To see this, note that a dictionary can only identify at most d distinct rays, but a cone, for example, may have many more rays. There is, however, a way to identify the lex-min basis for any given geometric ray.

Proposition 5.3. *The lexicographically minimum basis representing the geometric ray $y + tz$ is lex-positive.*

Proof. Let B be any basis representing the geometric ray $y + tz$, with dictionary given by (3.4). Let $s \in N$ index the column in this dictionary corresponding to z, and let $a = A_B^{-1} A_s$. We consider an *augmented* dictionary by adding a new variable x_q, $q = m + d + 1$, and appending to (3.4) the new equation:

$$x_s + x_q = 1.$$

This corresponds to adding the constraint $x_s \leq 1$ to the original problem. The augmented dictionary has basis $B+q$ and basic feasible solution $(x_B, x_q) = (A_B^{-1}b, 1)$. Column s no longer defines a ray in the augmented dictionary, and it is possible to pivot on this column with variable x_q leaving the basis, and x_s entering with value one. After the pivot the augmented system has basis $B + s$, basic feasible solution $(x_B, x_s) = (A_B^{-1}b - a, 1)$, and represents the *vertex* $y + z$ of the augmented problem. By Proposition 5.2, the lex-min basis $\overline{B}_{\min}$ for this vertex is lex-positive. Let $\overline{N}_{\min}$ be the corresponding cobasis. Since the basic feasible solution has $x_s = 1$, $s \in \overline{B}_{\min}$. Let $B_{\min} = \overline{B}_{\min} - s$. We have $q \in \overline{N}_{\min}$, since if not, x_q would be in the basis $\overline{B}_{\min}$ at value zero. As $q = m + d + 1$, it is the largest index, any degenerate pivot with q leaving the basis would produce a lexicographically smaller basis with the same basic feasible solution, contradicting the minimality of $\overline{B}_{\min}$. Since $q \in \overline{N}_{\min}$, the row in the dictionary for x_s is all zeroes except for a one in the columns for x_s and x_q.

We consider the pivot interchanging x_s and x_q in the augmented dictionary with basis $\overline{B}_{\min}$. Before the pivot, the column for x_q is $(a, 1)$, since increasing x_q to one changes the basic feasible solution from $(x_{B_{\min}}, x_q) = (A_B^{-1} b - a, 1)$ to $(x_{B_{\min}}, x_s) = (A_B^{-1} b, 1)$. The pivot preserves lex-positivity, since in the column for x_q only the last entry, corresponding to x_s, is positive. After the pivot the column for x_s is $(a, 1)$. If we delete the extra row from the augmented dictionary after the pivot, we get a dictionary for the original problem with lex-positive basis $B_{\min}$. It has basic feasible solution $x_{B_{\min}} = A_B^{-1} b$ and column s has value $-a$, so it represents the geometric ray $y + tz$. We claim that B is lexicographically at least as large as $B_{\min}$. This follows from the fact that $B + s$ is a basis for the vertex $y + z$ in the augmented problem, and so is lexicographically greater than or equal to the lex-min basis for this vertex, $\overline{B}_{\min} = B_{\min} + s$. $\qquad \square$

A basis representing a geometric ray can be tested for lex-minimality quite efficiently. The next proposition is analogous to Proposition 5.1; note that conditions (i)–(iii) are identical.

Proposition 5.4. *B is the lex-min basis representing a geometric ray $y + tz$ with basic feasible solution x if and only if there does not exist $r \in B$ and $s \in N$ such that*

(i) $r > s$,
(ii) $x_r = 0$,
(iii) $(A_B^{-1} A_s)_r \neq 0$, *and*
(iv) $(A_B^{-1} A_u)_r = 0$,

where $A_B^{-1} A_u$ represents the ray z.

Proof. The proof is similar to Proposition 5.1. If (i)–(iv) hold then the pivot interchanging r and s produces a smaller basis with the same properties. Conversely let $B_{\min}$ be the lex-min basis and B any other basis representing $y + tz$. Choose r and s as in Proposition 5.1. As shown there, (i)–(iii) must hold. It remains to observe that condition (iv) must hold or else the pivot would change column u by other than multiplication by a positive scalar, and the new basis would not represent $y + tz$. $\qquad \square$

The conditions of Proposition 5.4 can be tested in $O(md)$ time given the current dictionary. The results of this section are summarized in the following theorem.

Theorem 5.5. *Each vertex y and each geometric ray $y + tz$ of P can be represented uniquely by its lex-min basis, which is lex-positive.*

The goal of a vertex enumeration algorithm is to produce a minimum $\mathcal{V}$-representation (2.2) for P. In this minimum representation each direction z producing one or more extreme rays should be output once. When the polyhedron P is a pointed cone, the result of Theorem 5.5 is enough to achieve this. Since there is only one vertex, the origin, the only geometric ray with direction z is tz. In this case the set of lex-min bases gives a minimum $\mathcal{V}$-description of P. An important application of this is to facet enumeration of polytopes, see Section 6. For unbounded polyhedra with more than one vertex, we do not know any local necessary and sufficient condition to determine the lex-min basis for a ray (as opposed to a geometric ray). The following necessary condition allows some parallel geometric rays to be eliminated. Note conditions (i) and (iv) are the same as in Proposition 5.4.

Proposition 5.6. *B is the lex-min basis whose dictionary represents a ray z in column $A_B^{-1} A_u$ only if there does not exist $r \in B$ and $s \in N$ such that*

(i) $r > s$,

(ii) $(A_B^{-1} A_s)_r > 0$,

(iii) *r is an index which minimizes the ratio $(A_B^{-1} b)_i / (A_B^{-1} A_u)_i$, over all $i \in B$ for which the denominator is positive, and*

(iv) $(A_B^{-1} A_u)_r = 0$.

Proof. Conditions (i)–(iii) of the proposition imply that it is possible to make a feasible pivot to the basis $\overline{B} = B - r + s$, which is lexicographically smaller than B. Condition (iv) implies that the pivot does not change the column of the dictionary representing z. Let $\overline{x}$ be the basic feasible solution for basis $\overline{B}$. $\overline{B}$ is not necessarily lex-positive, but its dictionary represents the geometric ray $\overline{x} + tz$. By Proposition 5.3, the lex-min basis for this geometric ray is lex-positive. Since it is lexicographically no larger than $\overline{B}$, this proves B is not the lex-min basis of the ray z. $\qquad \square$

Testing the conditions of Proposition 5.6 requires $O(md)$ time, but is more expensive than testing those of Propositions 5.1 and 5.4 due to the additional ratio test required for condition (iii).

6. Reverse search

Based on Theorems 4.4 and 5.5 we can design pivoting algorithms to generate the vertices and extreme rays of P. An algorithm of this type is initiated with the lex-positive basis B^* and uses lexicographic pivoting to generate all lex-positive bases. A duplicate free list of vertices and geometric rays is obtained by outputting only the lex-min bases for each vertex and geometric ray. The algorithm can be described as a search of the graph whose nodes are lex-positive bases, and whose edges correspond to lexicographic pivots between these bases. A standard graph traversal algorithm, such as depth first search, could be used but suffers from the disadvantage that it is necessary to keep a list of all bases discovered. Even for rather small inputs, the size of such a list can be prohibitively large. *lrs* implements reverse search (see [2]), which is a method that enables the graph to be searched without maintaining a list of visited nodes. In the reverse search algorithm, only the current basis B, cobasis N and corresponding dictionary are stored.

The research search algorithm makes use of the following four functions.

(a) $pivot(B, r, s)$. B is the basis of the current dictionary, $r = B_i$ is a basic index and $s = N_j$ is a cobasic index. The dictionary is pivoted to the new basis $B - r + s$ as described in Section 3. The basic and cobasic indices B and N are updated by setting $B_i = s$ and $N_j = r$.

(b) $selectpivot(B, r, j)$. B is the basis of the current dictionary, which is assumed non-optimal. *selectpivot* returns indices $r \in B$ and $s \in N$ chosen by the lexicographic pivot rule applied to the dictionary with basis B. First the least index $s = N_j$ is found with negative coefficient in row zero, i. e., such than $(A_B^{-1} A_s)_0 < 0$. Then the lexicographic ratio test is performed on this column as described in Section 4, that is, we compute $r = lexminratio(B, s)$. The indices r and j are returned.

(c) $lexmin(B, s)$. B is the current basis representing a vertex y and s is either zero, or else $s \in N$ and the corresponding column of the dictionary for B represents a ray. When s is zero, *lexmin* determines if B is the lex-min basis for y, and if so returns *true*. Otherwise $s \in N$ and this column of the dictionary represents a ray z. *lexmin* determines if B is the lex-min basis for the geometric ray $y+tz$, and if so returns *true*. The operations required are given in Propositions 5.1 and 5.4.

(d) $reverse(B, u, v)$. Given basis B and cobasic index $v \in N$, *reverse* determines if there is a basic index $u \in B$ such that the lexicographic pivot rule applied to the dictionary with basis $\overline{B} = B - u + v$ generates a pivot back to B. In other words, *reverse* determines if there is an index $u \in B$ so that *selectpivot* applied to the dictionary with basis $B - u + v$ would compute the indices $r = v$ and $s = u$. This will happen if the first negative coefficient in row zero of the dictionary for $\overline{B}$ has index u, and if $v = lexminratio(\overline{B}, u)$. In this case, *reverse* returns *true* with the index u, otherwise it will return *false*. If the column v represents a ray, the ray is output if $lexmin(B, v)$ is *true*. *reverse* can be implemented by using *pivot* and *selectpivot*, but this is very inefficient. An efficient implementation is discussed below.

Using these four functions, the reverse search algorithm *lrs* can be described by the pseudo-code given in Figure 2. It is assumed that the dictionary with current basis B is available to all functions.

A general discussion of reverse search, proof of correctness and complexity analysis is given in [5], to which the reader is referred for more information. The main **while** loop is executed for each basis B, starting with the optimum basis. Each cobasic column of the current dictionary is examined by *reverse* to see if there is a lex-positive pivot using this column, for which the resulting dictionary pivots back to B using the lexicographic simplex method. If so the pivot is formed,

```
    B = B*; j = 1;
    repeat
          while j ≤ d
                {v = N_j;
                if reverse(B, u, v)
                then { pivot(B, u, v);                 // new basis found //
                            if lexmin(B, 0) then output current vertex;
                            j = 1;
                       }
                else j = j + 1;
                }
          selectpivot(B, r, j);              // backtrack //
          pivot(B, r, N_j);
              j = j + 1;
    until j > d and B = B*.
```

FIGURE 2. Pseudo-code for *lrs*

and the **while** loop is executed for each column of the new dictionary. When the **while** loop terminates for a given basis, *selectpivot* and *pivot* are used to return to its parent in the reverse search tree, and the **while** loop is continued for this basis. Note that for this it is essential that the value of j be correctly restored, as is done in *selectpivot*. The **until** statement is used to terminate the algorithm after the last column of the starting optimum dictionary has been examined.

Inspection of the pseudo-code shows that there are two pivots for each basis except the initial basis B^*: one in the **while** loop when a new basis is found, and one in the backtracking step, when a simplex pivot is used to move to the parent basis. Since pivoting is the most time consuming operation, a speedup can be achieved by saving the recent dictionaries in a cache, so they can be reloaded rather than recomputed. More critically, it is clear that *reverse* is executed d times for each basis and so must be implemented efficiently. This can be achieved by using the following proposition.

Proposition 6.1. *Given basis B, index $v \in N$, let $a = A_B^{-1} A_v$, and for any $t = 1, \ldots, m$ let w^t be the vector of coefficients of row t of the current dictionary* (3.4). *The function reverse(B, u, v) is true and returns $u = B_i$ if and only if*

(i) $w_v^0 < 0$,
(ii) $u = B_i = lexminratio(B, v) \neq 0$, *and*
(iii) *setting $\overline{w} = w^0 - a_0 w^i / a_i$, we have $\overline{w}_j \geq 0$, for all $j \in N, j < u$.*

Proof. Let $\overline{B} = B - u + v$. First suppose *reverse(B, u, v)* is *true*, which implies that *selectpivot* applied to the dictionary with basis $\overline{B}$ returns $r = v$ and $s = u$. This implies that $v = lexminratio(\overline{B}, u) \neq 0$. By Proposition 4.1(b), $u = lexminratio(B, v)$, giving (ii). It also implies that $a_i > 0$, and so $\overline{w}$ is well defined.

It follows from the pivot operation (Section 3) that $\overline{w}$ is the vector of coefficients of row zero of the dictionary with basis $\overline{B}$. Since *selectpivot* chooses $s = u$, $\overline{w}_u$ must be the first negative component of $\overline{w}$, so (iii) holds. Also by the pivot formula, $\overline{w}_u = w_v^0/a_i$. Since a_i is positive, (i) holds.

For the converse, assume that (i)–(iii) hold. Define $\overline{B}$ as above. As argued, $\overline{w}$ is the vector of coefficients of row zero of the dictionary corresponding to $\overline{B}$. From (ii) $a_i > 0$ and from (i) $w_v^0 < 0$, so $\overline{w}_u < 0$. Together with (iii) this implies that *selectpivot* applied to the dictionary with basis $\overline{B}$ will select u as the entering index. Since $u \neq 0$, we can apply Proposition 4.1(b), getting $v = lexminratio(\overline{B}, u)$. Therefore v is chosen as the leaving index by *selectpivot* and so $reverse(B, u, v)$ is true.   □

From Proposition 6.1 we see that *reverse* requires a lexicographic ratio test for each negative coefficient in row zero of the dictionary. If the ratio test succeeds, meaning the column does not represent an extreme ray, part of row zero of the updated dictionary is computed. The ratio test dominates the cost of this, requiring $O(md)$ time for degenerate dictionaries in the worst case. For non-degenerate dictionaries, the ratio test requires only $O(m)$ time, which since $m > d$ is also the time for testing all of the conditions (i)–(iii).

Referring to the pseudo-code in Figure 2, we see that the total time required by *reverse* for a basis B is $O(md^2)$ for degenerate dictionaries and $O(md)$ for non-degenerate dictionaries. The time required for an execution of *pivot*, *lexmin* or *selectpivot* is $O(md)$. This proves the following result.

Theorem 6.2. *lrs finds all bases and hence all vertices of a polyhedron in time $O(md^2)$ per basis and $O(md)$ space. It finds all vertices of a simple polyhedron in time $O(md)$ per vertex.*

7. Implementation issues and integer pivoting

To minimize space, *lrs* uses the reduced form of the dictionary (3.4) given by

$$x_B = A_B^{-1}b - A_B^{-1}A_N x_N \tag{7.1}$$

which is stored as a $(m+1) \times (d+1)$ array. Note that the signs of the coefficients of cobasic variables are the reverse of those in (3.4). Therefore the optimum dictionary is characterized by a row zero having non-positive coefficients. For any non-optimum dictionary, the candidates for entering variable for the simplex method are those with a positive coefficient in row zero. For definiteness, we choose the one with minimum index. For the lexicographic ratio test required to find the leaving variable, the rows of (4.1) required for (4.2) are recovered column by column, using Proposition 3.1, until a unique minimum is found. We can interpret lex-positivity in the context of the dictionary (3.4). B is a lex-positive basis if and only if for each $i \in B$ with $i \geq m+1$ and basic feasible solution $x_i = 0$, the first non-zero coefficient in its corresponding row of (7.1) is negative.

The basic and cobasic indices B and N are maintained in increasing order in *lrs*. In order to avoid moving the data in the dictionary, pointers are maintained to the actual row and column locations in the dictionary. This ordering allows certain operations to be optimized. For example, the test $B = B^*$ can be achieved by checking if $B_m = m$. Similarly the tests required in *selectpivot*, *lexmin* and *reverse* can be speeded up by processing the indices in order.

The reverse search method is extremely sensitive to numerical error. A single mistake in the sign of a dictionary element can mean that an entire subtree of the reverse search tree is not discovered. For that reason, exact arithmetic is used. *lrs* uses arrays to hold long integers, using the data format and basic routines given by Gonnet and Baeza-Yates [16]. The division routine is based on Knuth [19] and was implemented by Jerry Quinn.

The earlier version *rs* used rational arithmetic, and each entry in the dictionary was stored as a rational in reduced form, requiring *gcd* computations after each arithmetic operation. Most of these computations are eliminated in *lrs* by the use of integer pivoting method of Edmonds [15] (which is connected to Cramer's rule, see the appendix of Chvátal [13].) In integer pivoting, only the numerators of coefficients of the dictionary (7.1) are stored, with respect to a common denominator, which is the absolute value of the determinant of the current basis. The absolute value of the determinant is used so that the signs of the numerators agree with the signs of the rational numbers they represent. Let $a_{i,j}$, $0 \leq i \leq m$, $0 \leq j \leq d$ denote the array of coefficients of (7.1), and let $\det(B)$ be the determinant of the current basis B. If a pivot is to be performed on row r and column s, the updated (barred) coefficients are given by:

$$\bar{a}_{i,j} = (a_{i,j} a_{r,s} - a_{i,s} a_{r,j}) / \det(B)$$

$$\bar{a}_{r,j} = -a_{r,j}, \quad \bar{a}_{i,s} = a_{i,s}, \quad \bar{a}_{r,s} = \det(B), \quad \det(\bar{B}) = a_{r,s}$$

where in the above formulae $i \neq r$ and $j \neq s$. It can be shown that the integer division has no remainder. Since no *gcd* computations are required in integer pivoting, the only *gcd* operations performed in *lrs* are before printing the output. Empirically, integer pivoting appears to be between two and ten times faster than rational pivoting, with *gcd* computation using Euclid's algorithm.

8. Other functions of lrs

In this section we briefly described other functions of *lrs* that are built on top of the basic function of vertex enumeration. The details are given elsewhere, as noted.

8.1. Facet enumeration

The facet enumeration problem is to produce an $\mathcal{H}$-representation (2.1) of P from a $\mathcal{V}$-representation (2.2). In *lrs* a standard lifting technique (see for example Ziegler [20]) is used to convert the facet enumeration problem to an equivalent vertex enumeration problem. The input $\mathcal{V}$-representation is lifted to a pointed cone in one higher dimension, for which the two problems are equivalent. Specifically, each vertex $(a_1, \dots, a_d)$ of P is transformed to the inequality

$$x_1 + a_1 x_2 + \dots + a_d x_{d+1} \geq 0 \qquad (8.1)$$

and each ray $(c_1, \dots, a_d)$ of P is transformed to the inequality

$$a_1 x_2 + \dots + a_d x_{d+1} \geq 0.$$

The resulting system of inequalities describes a pointed cone $\overline{P}$ in $d+1$ dimensions. A ray $(z_1, \dots, z_{d+1})$ of $\overline{P}$ corresponds to the facet

$$z_1 + z_2 x_1 + \dots + z_{d+1} x_d \geq 0$$

of P. Note that no lifting is required if the input polyhedron P is a pointed cone. Also, if the input is the $\mathcal{V}$-representation of a polytope containing the origin, its $\mathcal{H}$-representation can be found without lifting by interpreting the input points as inequalities (8.1), with x_1 set to the constant one. Somewhat remarkably, although lifting increases degeneracy, *lrs* sometimes runs faster on lifted polytopes. This is due to the fact that rays are detected more efficiently than vertices, each of which require one (or two) additional pivots.

8.2. Voronoi vertices

Given a set of m points in $\mathbb{R}^d$, it is required to find the set of Voronoi vertices, each characterized by being the centre of an empty hypersphere spanned by at least $d+1$ input points. It is well known (see for example, Edelsbrunner [14]) that the Voronoi vertices of a set of points in $\mathbb{R}^d$ can be obtained by solving a vertex enumeration problem in $\mathbb{R}^{d+1}$. Each input point $(a_1, \dots, a_d)$ is transformed to the inequality

$$(a_1^2 + \dots + a_d^2) - 2a_1 x_1 - \dots - 2a_d x_d + x_{d+1} \geq 0.$$

The resulting system of inequalities describes a polyhedron $\overline{P}$ in $\mathbb{R}^{d+1}$. There is a one-to-one correspondence between the vertices of $\overline{P}$ and the Voronoi vertices of the input set of data. Indeed, each vertex $(y_1, \dots, y_{d+1})$ of $\overline{P}$ projects to the Voronoi vertex $(y_1, \dots, y_d)$. The inequalities described by the cobasic indices for a vertex of $\overline{P}$ also correspond to the input points defining the corresponding Voronoi vertex.

8.3. Volume computation

Given a set of m points in $\mathbb{R}^d$ it is required to find the volume of their convex hull. Let P be the polytope spanned by the input data points. As described in subsection 8.1, the facet enumeration of P can be obtained from the rays of the $d+1$-dimensional cone $\overline{P}$. It can be shown that the lex-positive bases of $\overline{P}$ form a decomposition of the polytope P into non-overlapping simplices. This follows from the fact that the lex-positive bases of $\overline{P}$ correspond to the vertices of a simple polyhedron obtained by a suitable perturbation of $\overline{P}$. The dual of this simple polyhedron is simplicial, and its projection gives the required decomposition of P. As we saw in Section 7, the determinant of each basis of $\overline{P}$ is readily available during the computation. The determinant is a equal to $d!$ times the volume of the corresponding simplex of P. Hence summing these determinants and dividing by $d!$ gives the required volume of P. See Büeler et al. [11] for more details and a comparison with other volume computation methods.

8.4. Estimation

The number of vertices in the $\mathcal{V}$-representation of a polyhedron represented by m inequalities in $\mathbb{R}^d$ can vary from zero to $m^{\lfloor d/2 \rfloor}$. P may have as many as m^d feasible bases, of which as many as $m^{\lfloor d/2 \rfloor}$ may be lex-positive.

Clearly, for even small values of m and d, if P achieves these upper bounds the vertex enumeration problem is intractable. As the running time of *lrs* is directly proportional to the number of bases computed, it is useful to estimate this number for any given polyhedron. Reverse search is amenable to a technique of tree estimation due to Hall and Knuth [17]. As described by Avis and Devroye [3], *lrs* can be used to estimate the number of lex-positive bases, and also the number of vertices and rays (or facets for facet enumeration problems). Similarly estimates for the volume of a polytope can be obtained. The estimates are unbiased, and techniques are given to lower the variance. In spite of the enormous range of the quantities to be estimated, the estimates obtained appear to give a good indication of the tractability of solving the given problem completely.

8.5. Restart capability

A feature of algorithms based on reverse search is that they can easily be interrupted and restarted in the middle of a computation. In the vertex enumeration setting, it is necessary to record only the indices of the current cobasis before interrupting the program. Then using this cobasis and the original input file, the current dictionary can be recomputed, and the computation resumed from this node of the reverse search tree. If cacheing is used, the cache is lost after the program is interrupted, but is restored automatically as new pivots are made after restarting.

9. Computational Results

In [4] some preliminary computational experience was given on a set of seven test problems. To illustrate the evolution of the code, we ran both the original program *ve01* and version 2.3 of *lrs* on these test problems, using the latest incarnation of *mutt*, a DEC AlphaServer 1000 4/23. The results are shown in Figure 3. In the table, $\#V$, $\#R$, $\#B$ refer respectively to the number of vertices, rays and bases computed. Note that due to lexicographic pivoting, *lrs* computes considerably fewer bases on the degenerate problems 2, 3 and 6. Problem 1 has a matrix generated uniformly in the range $-1000\ldots1000$ and b vector all ones. Problem 2 is due to Akihisa Tamura, and has all data from the set $\{0, -1, 1\}$. Problem 3 is the truncated metric cone on four points, consisting essentially of all triangle inequalities and non-negativity constraints on these four points. Problems 4 and 5 were constructed arbitrarily with integer data in the range $-100\ldots100$. Problem

6 was arbitrarily constructed with matrix entries in the range $-10 \ldots 10$ and b vector $1 \ldots 13$. Problem 7 is a Kuhn-Quandt problem, with matrix entries randomly chosen in the range $0 \ldots 1000$ and b vector entries all 10000. Its solution required integers with up to 63 decimal digits.

Problem	m	d	$\#V$	$\#R$	ve01		lrs	
					$\#B$	secs	$\#B$	secs
1	34	4	31		31	2.81	31	.06
2	16	5	18		1247	7.48	76	.05
3	19	6	8		10845	86.43	188	.12
4	12	7	54		54	.43	54	.05
5	14	9	89	33	97	1.14	94	.09
6	23	10	332	302	3656	83.65	824	1.69
7	20	10	1188		1188	208.98	1188	2.78

FIGURE 3. Computational results

One of the largest problems solved using *lrs* was a configuration polytope, arising in mathematical physics, with 96 inequalities in 86 dimensions. The output of 323,188 vertices was obtained by computing 1,621,760 bases in a computation lasting 30 days. Even larger problems have been solved using a parallel version of *lrs* described by Brüngger et al. in [10]. For example, they report computing the more than 3 million vertices and 57 million bases of a configuration polytope defined by 71 inequalities in 60 dimensions. More extensive computational experience with *lrs*, and an empirical analyses of the various speedups, is reported in Avis [7].

Acknowledgements. This paper was written while the author was on leave at IASI-CNR in Rome, whose support is gratefully acknowledged. Thanks to Elke Pose for the transformation of the paper to LATEX.
      Various people helped with aspects of the development of *lrs*, including David Bremner, Komei Fukuda, Ambros Marzetta, and Jerry Quinn. Finally, I would like to thank the users of *lrs* for suggestions, encouragement ... and for reporting bugs!

References

[1] lrs home page, May 1999, `ftp://mutt.cs.mcgill.ca /pub/C/lrs.html`

[2] D. Avis and K. Fukuda: A pivoting algorithm for convex hulls and vertex enumeration of arrangements and polyhedra, *Discrete Comput. Geometry* **8** (1992), 295–313.

[3] D. Avis and L. Devroye: Estimating the number of vertices of a polyhedron, in: "Snapshots of Computational and Discrete Geometry" (D. Avis and P. Bose, eds.), Vol. 3, School of Computer Science, McGill University, 1994, pp. 179–190, `ftp://mutt.cs.mcgill.ca/pub/doc/avis/AD94a.ps.gz`

[4] D. Avis: A C implementation of the reverse search vertex enumeration algorithm, in: *RIMS Kokyuroku* **872** (H. Imai, ed.), Kyoto University, May 1994, `ftp://mutt.cs.mcgill.ca /pub/doc/avis/Av94a.ps.gz`

[5] D. Avis and K. Fukuda: Reverse search for enumeration, *Discrete Applied Math.* **6** (1996), 21–46.

[6] D. Avis, D. Bremner, and R. Seidel: How good are convex hull algorithms?, *Comput. Geometry: Theory Appl.* **7** (1997), 265–301.

[7] D. Avis: Computational experience with the reverse search vertex enumeration algorithm, *Optimization Methods and Software* **10** (1998), 107–124.

[8] R. G. Bland: New finite pivoting rules for the simplex method, *Math. Operations Research* **2** (1977), 103–107.

[9] D. Bremner, K. Fukuda, and A. Marzetta: Primal-dual methods of vertex and facet enumeration, *Discrete Comput. Geometry* **20** (1997), 333–358.

[10] A. Brüngger, A. Marzetta, K. Fukuda, and J. Nievergelt: The parallel search bench ZRAM and its applications, 1997, `ftp://ftp.ifor.math.ethz.ch/pub/fukuda/reports`

[11] B. Büeler, A. Enge, and K. Fukuda: Exact volume computation for polytopes: A practical study, this volume, pp. 131–154.

[12] M. Bussieck and M. Luebbecke, The vertex set of a 0/1-polytope is strongly *P*-enumerable, *Comput. Geometry: Theory Appl.* **11** (1998), 103–109.

[13] V. Chvátal: Linear Programming, W.H. Freeman 1983.

[14] H. Edelsbrunner: Algorithms in Combinatorial Geometry, Springer-Verlag 1987.

[15] J. Edmonds and J.-F. Maurras, Note sur les *Q*-matrices d'Edmonds, *Recherche Opérationelle (RAIRO)* **31** (1997), 203–209.

[16] G. H. Gonnet and R. Baeza-Yates: Handbook of Algorithms and Data Structures, 2nd Edition, Addison-Wesley 1991.

[17] M. Hall and D. E. Knuth, Combinatorial analysis and computers, *Amer. Math. Monthly* **72** (1965), 21–28.

[18] J. Ignizio and T. Cavalier: Linear Programming, Prentice Hall 1994.

[19] D. E. Knuth: The Art of Computer Programming, Vol 2: Seminumerical Algorithms, Addison-Wesley 1981; Third edition 1998.

[20] G. M. Ziegler: Lectures on Polytopes, Graduate Texts in Mathematics **152**, Springer-Verlag 1994, revised 1998.

David Avis
School of Computer Science
McGill University
3480 University
Montréal, Québec
Canada H3A 2A7
`avis@cs.mcgill.ca`

The Complexity of Yamnitsky and Levin's Simplices Algorithm

Sven G. Bartels

Abstract. In 1982 Yamnitsky and Levin gave a variant of the ellipsoid method which uses simplices instead of ellipsoids. Unlike the ellipsoid method this simplices method can be implemented in rational arithmetic. We show, however, that this results in a non-polynomial method since the storage requirement may grow exponentially with the size of the input. Nevertheless, by introducing a rounding procedure we can guarantee polynomiality for both a central-cut and a shallow-cut version. Thus in most applications the simplices method can serve as a substitute for the ellipsoid method. In particular, it performs better than the ellipsoid method if it is used to obtain bounds for the volume of a convex body. Furthermore, it can be used to estimate the optimal function value of total approximation problems.

1. Introduction

In 1979 L.G. Khachian introduced the ellipsoid method which initiated a flood of further research in the fields of linear programming, nonlinear optimization, and combinatorial optimization [9], [11], [16]. In its basic form, this method either determines a feasible point of the convex polyhedron

$$P := \{x \in \mathbb{R}^n | Ax \le b\}$$

with integer data $A \in \mathbb{Z}^{m \times n}, b \in \mathbb{Z}^m$, or reports that the interior of P is empty. Assuming that all arithmetic operations are carried out to infinite precision, it is relatively simple to show that the method needs a polynomial number of these operations in terms of the digital size of the system of linear inequalities $L := \sum_{i=1}^{m} \sum_{j=1}^{n} \langle A_{ij} \rangle + \sum_{i=1}^{m} \langle b_i \rangle$, where $\langle a \rangle := \lceil \log_2(|a| + 1) \rceil + 1$ for all $a \in \mathbb{Z}$. Since the ellipsoid method involves taking square roots, this assumption soon becomes a major obstacle. Using a particular type of rounding procedure, it is possible to obtain an ellipsoid method such that the total number of bit operations is polynomially bounded (see [5], [9], [16]). The corresponding proofs require, however, considerable technical effort.

Edited by: Volker Kaibel, MA 7–1, Fachbereich Mathematik, TU Berlin, D-10623 Berlin, Germany, kaibel@math.TU-Berlin.DE.

In 1982 B. Yamnitsky and L.A. Levin gave a variant of the ellipsoid method which solves the same problem, but uses simplices instead of ellipsoids [6], [22]. For the convenience of the reader, this simplices method is described in the next section. Like the original ellipsoid method it needs a polynomial number of arithmetic operations, but this time only the four standard arithmetic operations are used. By using rational arithmetic, it is therefore possible to carry out all arithmetic operations to infinite precision, and thus one might hope that with this method there is no need for rounding. Indeed this has been proposed in [1]. However, in Section 3 an example is given that shows that precise rational arithmetic may result in an exponential output size. Therefore a rounding procedure is still required which is given in Section 4. On the positive side, we feel that the proofs of its properties are somewhat simpler than in the case of the ellipsoid method. Section 5 deals with the modifications necessary to obtain a shallow-cut version of the algorithm. This yields methods for estimating the volume of a convex body and for obtaining bounds for the optimal function value of a concave minimization problem which are discussed in the last section.

2. The basic algorithm

In the following we give an outline of the simplices method without rounding. It is based on the presentation of the algorithm by V. Chvátal in [6]. The first difference to the original algorithm in [22] lies in the presentation of the simplices just by their vertices. The second difference is the use of a central-cut version instead of a deep-cut version of the method.

For conceptual reasons the algorithm is given for the following nonemptiness problem for a general compact convex set $K \subset \mathbb{R}^n$.

Nonemptiness problem

Input: A positive rational number ε.

Output: Either a vector $c \in K$, or the assertion that $\mathrm{vol}(K) \leq \varepsilon$.

In order to solve this problem, we assume that we are given an oracle that can solve the following problem:

Separation problem

Input: A vector $c \in \mathbb{R}^n$.

Output: Either the assertion that $c \in K$, or a vector $a \in \mathbb{R}^n \backslash \{0\}$
                such that $K \subset \{x \in \mathbb{R}^n | a^T x < a^T c\}$.

An oracle is simply a subroutine that can produce a certain output (see [9]). There is no assumption made on how the subroutine finds a solution. The existence of the separating hyperplane for $c \notin \mathrm{int}(K)$ follows from a standard separation theorem (see, e.g., [18] Theorem 11.3).

Given an oracle for the separation problem, the simplices method below can now be used to solve the nonemptiness problem. Some further explanatory remarks and an example will be given afterwards.

Simplices method

Input: Affinely independent vectors $v_0^{[0]}, \ldots, v_n^{[0]} \in \mathbb{Q}^n$ such that

$$K \subseteq S^{[0]} := \mathrm{conv}\{v_0^{[0]} \ldots, v_n^{[0]}\},$$

a rational positive number ε, and an integral positive number N such that

$$N \geq 2(n+1)^2 \ln \left(\frac{\mathrm{vol}(S^{[0]})}{\varepsilon} \right).$$

Step 1:   Set $k := 0$.

Step 2:   Determine the centroid of $S^{[k]}$

$$c^{[k]} := \frac{1}{n+1} \sum_{i=0}^{n} v_i^{[k]}.$$

Step 3:   Call the oracle for the separation problem with the vector $c^{[k]}$. If the oracle asserts that $c^{[k]} \in K$, then STOP (a feasible point is found).

Step 4:   Determine an index $l \in \{0, \ldots, n\}$ such that

$$a^T v_l^{[k]} = \min_{i \in \{0, \ldots, n\}} a^T v_i^{[k]}.$$

Step 5:   Determine a new simplex $S^{[k+1]} := \mathrm{conv}\{v_0^{[k+1]}, \ldots, v_n^{[k+1]}\}$ given by

$$v_i^{[k+1]} := (1 - \delta_i) v_l^{[k]} + \delta_i v_i^{[k]},$$

where

$$\delta_i := \left(1 - \frac{a^T(c^{[k]} - v_i^{[k]})}{n^2 a^T(c^{[k]} - v_l^{[k]})} \right)^{-1}.$$

Step 6:   If $k = N$, then STOP ($\mathrm{vol}(K) < \varepsilon$), otherwise set $k := k + 1$ and go to Step 2.

It is easily seen that for every $i \in \{0, \ldots, n\} \setminus \{l\}$ the new vertex $v_i^{[k+1]}$ obtained in Step 5 lies on the halfline from $v_l^{[k]}$ through $v_i^{[k]}$. For the following discussion it is convenient to introduce the affine function

$$e : \mathbb{R}^n \to \mathbb{R}, \quad e(x) := a^T(c^{[k]} - x).$$

The term for δ_i in Step 5 can then be written as

$$\delta_i := \left(1 - \frac{e(v_i^{[k]})}{n^2 e(v_l^{[k]})} \right)^{-1}.$$

The inequality $e(x) \geq 0$ defines a halfspace which contains K and whose boundary contains the centroid $c^{[k]}$. In [6] and [22] it is proved that

$$S^{[k]} \cap \{x \in \mathbb{R}^n \mid e(x) \geq 0\} \subseteq S^{[k+1]} \tag{1}$$

and that

$$\frac{\text{vol}(S^{[k+1]})}{\text{vol}(S^{[k]})} \le e^{-1/(2(n+1)^2)}. \tag{2}$$

The inequalities (1) and (2) still hold if the function $e : \mathbb{R}^n \to \mathbb{R}$ is chosen as $e(x) := \beta - a^T x$, where $\beta < a^T c^{[k]}$ and $K \subset \{x \in \mathbb{R}^n \mid a^T x \le \beta\}$. In the terminology of the ellipsoid method, this corresponds to a deep-cut instead of a central-cut version of the algorithm. In either case their might exist simplices $S^{[k+1]}$ which satisfy (1) and which have a smaller volume than the new simplex given by Step 5. A simple improvement in either version is to set

$$\delta_i := \left(1 - \frac{e(v_i^{[k]})}{r^2 e(v_i^{[k]})}\right)^{-1}, \tag{3}$$

where

$$r := \#\{i \in \{0, \dots, n\} \mid e(v_i^{[k]}) > 0\}.$$

An update formula which decreases the volume even further is given by R. Prakash and K.J. Supowit [17]. This however requires the minimization of a strictly convex one-dimensional function by some numerical method. In a practical implementation, these modifications are likely to improve the performance of the simplices method. Indeed some encouraging numerical results for the Prakash-Supowit version are given in [8]. Here we nevertheless chose the central-cut version given above since its complexity is somewhat easier to analyse.

In order to use this algorithm for finding a feasible point of a convex polyhedron $P := \{x \in \mathbb{R}^n \mid Ax \le b\}$, we need two well-known lemmas. Using Cramer's Rule, it can be shown that

$$\text{int}(P \cap \{x \in \mathbb{R}^n \mid \|x\|_\infty \le 2^L\}) \ne \emptyset \tag{4}$$

unless $\text{int}(P) = \emptyset$ (see, e.g., [9] Lemma 3.1.33). In order to decide whether the interior of P is empty, it is thus sufficient to consider the restricted polytope given by (4). The vertices of the simplex required can be obtained by choosing the vertices

$$\begin{aligned} v_0^{[0]} &:= 2^L[-1, \dots, -1]^T, \\ v_i^{[0]} &:= v_0^{[0]} + 2^{L+1} n e_i \quad \text{for } i \in \{1, \dots, n\}, \end{aligned} \tag{5}$$

where $e_i \in \mathbb{R}^n$ denotes the i-th unit vector. The volume of the initial simplex is given by

$$\text{vol}(S^{[0]}) = \frac{n^n}{n!} 2^{(L+1)n}. \tag{6}$$

Furthermore it can be shown that if $\text{int}(P) \ne \emptyset$, then $\text{vol}(P) > 2^{-(n+1)L}$ (see [9] Lemma 3.1.35). In order to decide whether the interior of P is empty, it is therefore appropriate to choose $\varepsilon := 2^{-(n+1)L}$. The oracle for the separation problem typically chooses an inequality $a^T x \le \beta$ that is contained in the system of linear inequalities describing P and that is violated by the centroid $c^{[k]}$. This can obviously be done by simply checking each inequality. For specific polytopes it might be possible to find out an inequality like this in a more efficient way. This is the key for deriving polynomial algorithms for certain combinatorial problems which lead to polytopes with an exponential number of facets (see [9]).

The geometric idea of the method is illustrated by the following example. The polytope

$$K = P := \{x \in \mathbb{R}^2 \mid -1/2 \le x_1 \le 1/2, \ -7/8 \le x_2 \le -1/2\} \tag{7}$$

is contained in the simplex

$$S^{[k]} = \text{conv}\{[-1, -1]^T, [0, 2]^T, [1, -1]^T\}.$$

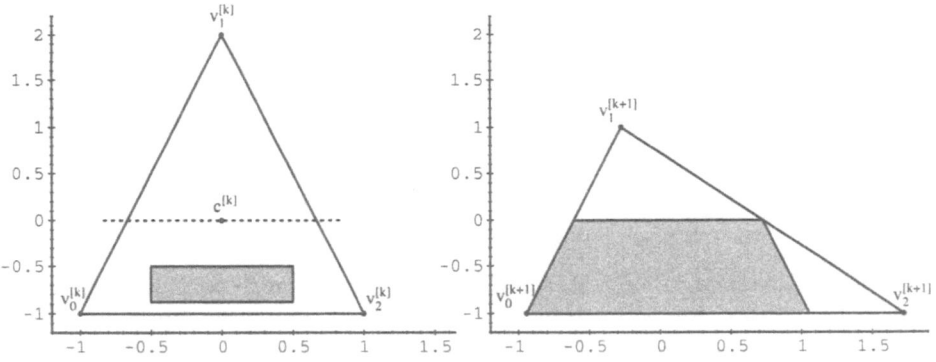

FIGURE 1. A central cut

Only $x_2 \le -1/2$ is violated by the centroid $c^{[k]} = [0,0]^T$, therefore a separating hyperplane is given by $a = [0,1]^T$. Since

$$a^T v_0^{[k]} = a^T v_2^{[k]} = \min_{i \in \{0,1,2\}} a^T v_i^{[k]},$$

either of these two points may be chosen as $v_l^{[k+1]}$. The second figure shows the new simplex generated in Step 5.

3. An example for exponential output size

Let $K = P := \{x \in \mathbb{R}^2 \mid Ax \le b\}$ be given by

$$A := \begin{bmatrix} -1 & 0 \\ 0 & -1 \\ 2^s & 2^s \end{bmatrix} \quad b := \begin{bmatrix} 0 \\ 0 \\ 1 \end{bmatrix}$$

with $s \in \mathbb{N} \cup \{0\}$, so the digital size of the system is $L = 2s + 14$. Let the initial simplex $S^{[0]}$ be given by the vertices

$$v_0^{[0]} := [0,0]^T, \quad v_1^{[0]} := [d,0]^T, \quad v_2^{[0]} := [0,1]^T$$

with $d \in \mathbb{N} \setminus \{1\}$ fixed. Furthermore it is assumed that if the input vector is not feasible, the oracle for the separation problem returns the coefficient vector from a restriction which is violated by the input. The following table shows the performance of the simplices algorithm for this example for $d = 2$.

s	# iter.	# digits
0	0	7
1	6	484
2	12	28522

The second column of the table gives the number of the iterations in which a feasible centroid is found (e.g. for $s = 0$ the initial centroid $c^{[0]}$ is feasible). The third column gives the number of digits which are needed to denote the coordinates of the final centroid given by fractions of binary numbers. By definition one has to add four in order to obtain the digital size of the output. This example shows a rapid increase of the number of digits of the output. By the following theorem, this number grows exponentially with L.

Theorem 1. *For $d \geq 5$ in the given example the simplices method generates a feasible point with a digital size that grows exponentially with respect to L.*

Before giving a proof, we state the main consequence of the example. Although for convex polyhedra the number of iterations of the simplices method does depend polynomially on L, the storage requirement and therefore the number of bit operations does not. In order to obtain a polynomial algorithm, one needs some additional rounding procedure as given in the next section. Note however that this example depends on the version of the simplices method. In fact, with the modification (3) from the previous section the simplices method finds a feasible point within one iteration for every $s \in \mathbb{N}$. However, we have numerical evidence that with other examples one experiences an exponential growth of the digital size with this modification, too.

The proof of this theorem is divided up into six lemmas and a final main proof.

Lemma 1. *For the given example, the vertices generated can be described by*

$$v_0^{[k]} = [0,0]^T, \quad v_1^{[k]} = [x^{[k]}, 0]^T, \quad v_2^{[k]} = [0, y^{[k]}]$$

with

$$x^{[k]} := \left(\frac{4}{3}\right)^k d \frac{\alpha_k(d)}{\beta_k(d)}, \quad y^{[k]} := \left(\frac{4}{3}\right)^k \frac{\alpha_k(d)}{\gamma_k(d)},$$

where

$$\alpha_k(d) := \Pi_{i=0}^k p_i(d), \quad \beta_k(d) := \Pi_{i=0}^k q_i(d), \quad \gamma_k(d) := \Pi_{i=0}^k r_i(d),$$
$$p_{k+1}(d) := d\gamma_k(d) + \beta_k(d), \quad q_{k+1}(d) := 2d\gamma_k(d) + \beta_k(d),$$

$$r_{k+1}(d) := d\gamma_k(d) + 2\beta_k(d), \quad p_0(d) = q_0(d) = r_0(d) := 1.$$

Proof. For $k = 0$, we have $x^{[0]} = d$ and $y^{[0]} = 1$ as required. Now assume that $v_0^{[k]}, v_1^{[k]}, v_2^{[k]}$ with $k \in \mathbb{N} \cup \{0\}$ fixed are given as above. The centroid of the simplex is then given as

$$c^{[k]} := \frac{1}{3}[x^{[k]}, y^{[k]}]^T.$$

In the next iteration, the simplices method generates the points

$$
\begin{aligned}
v_0^{[k+1]} &:= v_0^{[k]}, \\
v_1^{[k+1]} &:= (1 - \delta_1)v_0^{[k]} + \delta_1 v_1^{[k]}, \\
v_2^{[k+1]} &:= (1 - \delta_2)v_0^{[k]} + \delta_2 v_2^{[k]},
\end{aligned}
$$

with

$$\delta_1 := \left(1 - \frac{(x^{[k]} + y^{[k]})/3 - x^{[k]}}{4(x^{[k]} + y^{[k]})/3}\right)^{-1} = \frac{4}{3}\frac{x^{[k]} + y^{[k]}}{2x^{[k]} + y^{[k]}},$$

$$\delta_2 := \left(1 - \frac{(x^{[k]} + y^{[k]})/3 - y^{[k]}}{4(x^{[k]} + y^{[k]})/3}\right)^{-1} = \frac{4}{3}\frac{x^{[k]} + y^{[k]}}{x^{[k]} + 2y^{[k]}}.$$

Thus the points of the next iteration are given as $v_0^{[k+1]} = [0,0]^T$ and $v_1^{[k+1]} = [x^{[k+1]}, 0]^T$ with

$$
\begin{aligned}
x^{[k+1]} &= \frac{4}{3}\frac{x^{[k]} + y^{[k]}}{2x^{[k]} + y^{[k]}}x^{[k]} \\
&= \frac{4}{3}\left(\frac{4}{3}\right)^k d\,\frac{\alpha_k(d)}{\beta_k(d)}\,\frac{(4/3)^k d\alpha_k(d)/\beta_k(d) + (4/3)^k \alpha_k(d)/\gamma_k(d)}{2(4/3)^k d\alpha_k(d)/\beta_k(d) + (4/3)^k \alpha_k(d)/\gamma_k(d)} \\
&= \left(\frac{4}{3}\right)^{k+1} d\,\frac{\alpha_k(d)}{\beta_k(d)}\,\frac{d\gamma_k(d) + \beta_k(d)}{2d\gamma_k(d) + \beta_k(d)} = \left(\frac{4}{3}\right)^{k+1} d\,\frac{\alpha_{k+1}(d)}{\beta_{k+1}(d)}.
\end{aligned}
$$

The identity for $v_2^{[k+1]}$ can be obtained in the same way. □

Lemma 2. *Let $\alpha_k(d)$, $\beta_k(d)$, $\gamma_k(d)$ be given as above. Then*

$$\alpha_k(d) \geq d^{2^{k-1}}, \quad \beta_k(d) \geq (2d)^{2^{k-1}}, \quad \gamma_k(d) \geq d^{2^{k-1}}.$$

Proof. Together with

$$p_{k+1}(d) \geq d^{2^{k-1}}, \quad q_{k+1}(d) \geq (2d)^{2^{k-1}}, \quad r_{k+1}(d) \geq d^{2^{k-1}},$$

the proof can easily be carried out by induction. □

Although the previous lemma gives lower bounds for the terms involved in $x^{[k]}$ and $y^{[k]}$, it does not take into account any cancellations which might be possible. In the following the objective is therefore to get an upper bound for the greatest common divisor of the numerator and the denominator of $x^{[k]}$. Some technical identities are summarized in the following lemma.

Lemma 3. *Let $k \geq 1$. Then*

(i) $3p_k(d) = q_k(d) + r_k(d)$,
(ii) $q_k(d) - p_k(d) = d\gamma_{k-1}(d)$,
(iii) $2p_k(d) - q_k(d) = \beta_{k-1}(d)$,
(iv) $2q_k(d) - 3p_k(d) = (d-1)\alpha_{k-1}(d)$.

Proof. The identities (i) to (iii) immediately follow from the definitions given in (Lemma 1). The fourth identity is proved by induction. For $k = 1$, we have

$$2q_1(d) - 3p_1(d) = 2(2d+1) - 3(d+1) = d - 1 = (d-1)\alpha_0(d).$$

Now assume that $2q_k(d) - 3p_k(d) = (d-1)\alpha_{k-1}(d)$ for some fixed $k \in \mathbb{N}$. Then

$$
\begin{aligned}
2q_{k+1}(d) - 3p_{k+1}(d) &= d\gamma_k(d) - \beta_k(d) \\
&= d\gamma_{k-1}(d)r_k(d) - \beta_{k-1}(d)q_k(d) \\
&= d\gamma_{k-1}(d)(d\gamma_{k-1}(d) + 2\beta_{k-1}(d)) \\
&\quad -\beta_{k-1}(d)(2d\gamma_{k-1}(d) + \beta_{k-1}(d)) \\
&= (d\gamma_{k-1}(d) + \beta_{k-1}(d))(d\gamma_{k-1}(d) - \beta_{k-1}(d)) \\
&= p_k(d)(2q_k(d) - 3p_k(d)) \\
&= (d-1)\alpha_k(d).
\end{aligned}
$$

The assumption of the induction has only been used for the last equality. $\square$

The following lemma states important identities for the greatest common divisors (gcd) of the multiples of the fraction above.

Lemma 4. *Let $p_k(d)$, $q_k(d)$, $r_k(d)$ be given as above. Then*

$$
\begin{aligned}
\gcd(p_k(d), q_k(d)) &= \gcd(p_k(d), \beta_{k-1}(d)) \\
&= \gcd((d-1)\alpha_{k-1}(d), \beta_{k-1}(d)), \\
\gcd((d-1)\alpha_{k-1}(d), q_k(d)) &= \gcd(3\beta_{k-1}(d), (d-1)\alpha_{k-1}(d)).
\end{aligned}
$$

Proof. The above identities can be shown by using that

$$\gcd(a,b) = \gcd(\min\{a,b\}, |a-b|)$$

for $a, b \in \mathbb{N}, a \neq b$. Hence

$$
\begin{aligned}
\gcd(p_k(d), q_k(d)) &= \gcd(p_k(d), q_k(d) - p_k(d)) \\
&= \gcd(q_k(d) - p_k(d), 2p_k(d) - q_k(d)) \\
&= \gcd(2p_k(d) - q_k(d), 2q_k(d) - 3p_k(d)) \\
&= \gcd(\beta_{k-1}(d), (d-1)\alpha_{k-1}(d)).
\end{aligned}
$$

Since by Lemma 3 all the terms involved are positive, the above tranformations are valid. The remaining identities can be derived in the same way. $\square$

With the following final two lemmas we provide lower bounds for the numerator and the denominator of $x^{[k]}$.

Lemma 5. *Let* $k \geq 1$. *Then* 3^{2^k-1} *can be divided by* $\gcd((d-1)\alpha_k(d), \beta_k(d))$.

Proof. For $k = 1$, we have that

$$\gcd((d-1)\alpha_1(d), \beta_1(d)) = \gcd((d-1)(d+1), 2d+1)$$

is a divisor of

$$\gcd(d-1, 2d+1)\gcd(d+1, 2d+1) = \gcd(d-1, 3)\gcd(d, 1),$$

which in turn is divisor of $3 = 3^{2^1-1}$. Now assume that the above statement holds for some fixed k. We then have

$$
\begin{aligned}
& \gcd((d-1)\alpha_{k+1}(d), \beta_{k+1}) \\
= \; & \gcd((d-1)\alpha_k(d)p_{k+1}(d), \beta_k(d)q_{k+1}(d)) \\
= \; & \gcd((d-1)\alpha_k(d), \beta_k(d))\gcd(p_{k+1}(d), q_{k+1}(d)) \\
& \times \gcd\left(\frac{(d-1)\alpha_k(d)}{\gcd((d-1)\alpha_k(d), \beta_k(d))}, \frac{q_{k+1}(d)}{\gcd(p_{k+1}(d), q_{k+1}(d))}\right) \\
& \times \gcd\left(\frac{\beta_k(d)}{\gcd((d-1)\alpha_k(d), \beta_k(d))}, \frac{p_{k+1}(d)}{\gcd(p_{k+1}(d), q_{k+1}(d))}\right).
\end{aligned}
$$

By Lemma 4, this shows that

$$\gcd((d-1)\alpha_{k+1}(d), \beta_{k+1}(d))$$

is a divisor of

$$
\begin{aligned}
& \gcd((d-1)\alpha_k(d), \beta_k(d))^2 \\
& \times \gcd\left(\frac{3(d-1)\alpha_k(d)}{\gcd(3\beta_k(d), (d-1)\alpha_k(d))}, \frac{3q_{k+1}(d)}{\gcd(3\beta_k(d), (d-1)\alpha_k(d))}\right) \\
& \times \gcd\left(\frac{\beta_k(d)}{\gcd(\beta_k(d), P_{k+1}(d))}, \frac{p_{k+1}(d)}{\gcd(\beta_k(d), p_{k+1}(d))}\right),
\end{aligned}
$$

and therefore (by the induction hypothesis for the first factor and by Lemma 4 for the second one) a divisor of

$$3^{2(2^k-1)} \times 3 \times 1 = 3^{2^{(k+1)}-1}.$$

The result follows.   □

Lemma 6. *Let* $k \geq 1$ *and* $x^{[k]} = a_k/b_k$ *with* $a_k, b_k \in \mathbb{N}$. *Then*

$$a_k \geq \left\lceil \frac{1}{3^k}\left(\frac{d}{9}\right)^{2^{k-1}} \right\rceil, \quad b_k \geq \left\lceil \frac{1}{4^k d}\left(\frac{2d}{9}\right)^{2^{k-1}} \right\rceil.$$

Proof. By Lemma 1 the vector $x^{[k]}$ can be represented as

$$x^{[k]} = \left(\frac{4}{3}\right)^k \frac{d}{d-1} \frac{(d-1)\alpha_k(d)}{\beta_k(d)}.$$

By Lemma 2 the numerator of the above fraction is greater or equal to

$$4^k d(d-1)d^{2^{k-1}}.$$

By Lemma 5 the numerator and the denominator can be cancelled by at most

$$12^k d(d-1)3^{2^k-1}.$$

Hence the result for a_k follows. The lower bound for b_k can be derived in the same way.   □

Proof of Theorem 1. By the Lemmas 1 and 3 we have

$$x^{[k]} = \left(\frac{4}{3}\right)^k d \; \frac{\Pi_{i=0}^k((q_i(d)+r_i(d))/3)}{\Pi_{i=0}^k q_i(d)} > \left(\frac{4}{9}\right)^k \frac{d}{3}.$$

The algorithm stops if the centroid of the current simplex is feasible. Therefore the number of iterations can be bounded by

$$\left(\frac{4}{9}\right)^k \frac{d}{9} < 2^{-s} \Leftrightarrow k > \frac{s+\log_2(d)-\log_2(9)}{\log_2(9)-2} = \frac{(L-14)/2+\log_2(d)-\log_2(9)}{2\log_2(9)-2},$$

or simpler by

$$k > 2L/5 - 9.$$

Together with Lemma 6 this implies that the denominator of $x^{[k]}$ (after termination) grows (with increasing s) at least as fast as

$$\text{const} \cdot \left(\frac{10}{9}\right)^{2^{\text{const} \cdot L}}.$$

The result follows.   □

4. Using rounding to obtain a polynomial algorithm

In the previous section it has been demonstrated that exact arithmetic might result in an exponential storage requirement. This problem can be cured with the modified method presented below. It yields a polynomial time-method for solving the nonemptiness problem for convex polyhedra.

Theorem 2. *Consider the following modification of Steps 5 and 6 of the simplices method:*

Simplices method with rounding

Input: *Affinely independent vectors* $v_0^{[0]}, \ldots, v_n^{[0]} \in \mathbb{Q}^n$ *such that*

$$K \subseteq S^{[0]} := \mathrm{conv}\{v_0^{[0]} \ldots, v_n^{[0]}\},$$

a rational positive number ε, *and an integral positive number* N *such that*

$$N \geq 3(n+1)^2 \ln \left(\frac{\mathrm{vol}(S^{[0]})}{\varepsilon} \right).$$

Step 5:   *Determine a new simplex* $S^{[k+1]} := \mathrm{conv}\{v_0^{[k+1]}, \ldots, v_n^{[k+1]}\}$ *given by*

$$v_i^{[k+1]} := (1 - \bar{\delta}_i)v_l^{[k]} + \bar{\delta}_i v_i^{[k]},$$

where

$$\bar{\delta}_i := 2^{-t} \lceil \delta_i 2^t \rceil,$$

$$\delta_i := \left(1 - \frac{a^T (c^{[k]} - v_i^{[k]})}{n^2 a^T (c^{[k]} - v_l^{[k]})} \right)^{-1},$$

$$t := 3\langle n \rangle.$$

Then the modified algorithm still gives a correct result. If the components of $v_0^{[0]}, \ldots, v_n^{[0]}$ *are integral, then the storage requirement for* $v_0^{[k]}, \ldots, v_n^{[k]}$ *is bounded from above by*

$$n(n+1) \left(3N\langle n \rangle + \max_{j \in \{0,\ldots,n\}} \langle \|v_j^{[0]}\|_\infty \rangle \right)$$

bits.

The proof will be given in greater generality in the next section. The main idea of the modification lies in the fact that the parameters δ_i are **rounded up** to t digits after the decimal point to give a new approximate parameter $\bar{\delta}_i$. With these new parameters, the new vertices are then computed **exactly**. Since $\bar{\delta}_i \geq \delta_i$, it is guaranteed that the resulting new simplex contains the simplex that would be generated by the original method without rounding.

In each iteration the maximum number of digits behind the decimal point of the components $v_0^{[k]}, \ldots, v_n^{[k]}$ increases by at most t digits. Assuming that the components $v_0^{[0]}, \ldots, v_n^{[0]}$ have integer values, the number of digits behind the decimal point of the components of $v_0^{[k]}, \ldots, v_n^{[k]}$ is bounded from above by kt. Together with an upper bound of $\|v_i^{[k]}\|_\infty$ (see Lemma 11), this yields the upper bound for the storage requirement of the vector $v_0^{[0]}, \ldots, v_n^{[k]}$.

The geometric idea is again illustrated by the example (7). Figure 2 shows the new simplices that are obtained by the original method and by the method with rounding. In order to give a better visualisation of the difference, we chose $t = 3$ instead of $t = 6$ as given by the algorithm.

The following corollary states that a common prototype problem can be solved in polynomal time.

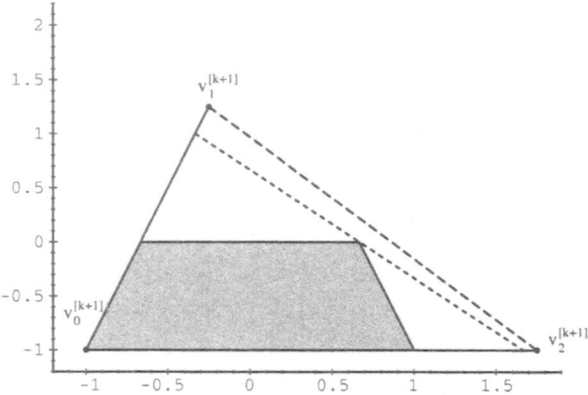

FIGURE 2. A central cut with rounding

Corollary 1. *Let P be a convex polytope such that there exist positive integral numbers r and R such that*

$$\{x \in \mathbb{R}^n \mid \|x - x_0\|_\infty \le 1/r\} \subseteq P \subseteq \{x \in \mathbb{R}^n \mid \|x\|_\infty \le R\}$$

for some $x_0 \in \mathbb{R}^n$. Assume that an oracle for the separation problem is given that, if it finds a separating hyperplane, returns a vector $a \in \mathbb{Q}^n$ with a digital size that is bounded by some variable ϕ. Then if the simplices method with rounding is called with the input

$$
\begin{aligned}
v_0^{[0]} &:= R[-1, \dots, -1]^t, \\
v_i^{[0]} &:= v_0^{[0]} + 2nR, \quad i \in \{1, \dots, n\}, \\
\varepsilon &:= (2/r)^n, \\
N &:= 3n(n+1)^2(\langle R \rangle + \langle n \rangle + \langle r \rangle),
\end{aligned}
$$

where e_i denotes the i-th unit vector, the method stops in Step 3 with a feasible point. The algorithm is polynomial in $n, \langle R \rangle, \langle r \rangle, \phi$, and the number of oracle calls.

Proof. It is easily checked that the input delivers the correct result. The dependence of the complexity on ϕ stems from the computations of $\bar{\delta}_i$. □

It follows from the discussion given in Section 2 that the nonemptiness problem for convex polyhedra can be solved by the simplices method with rounding with a number of bit operations that is polynomial in L.

At this point we mention another obstacle which stems from the general formulation of the simplices method for arbitrary convex sets. For example, the set K might be given in a form which makes it impossible to decide whether some point on the boundary of K is an element of K or not. Also, if the oracle for the separation problem returns a separating hyperplane given by a vector $a \in \mathbb{R}^n$ with irrational coordinates, then usually some kind of rational approximation will be needed which leads to further rounding errors. In order to account for such problems, the so-called weak separation problem was introduced (see [9]). Although it should be possible to develop a simplices method based on this weak version like in the case of the ellipsoid method, we sticked to the strong version for the sake of simplicity. It is sufficient for many important applications.

5. A shallow-cut version

In this section a shallow-cut version of the simplices method is derived. Applications of the method are considered in Section 6.

If the simplices method stops in Step 3, the volume of the last simplex is obviously bounded by the minimal volume of all simplices containing the set K. An interesting property of this simplex resembling a corresponding property of the Löwner-John-ellipsoid (see [9] Theorem 3.1.9) is given by the following theorem. For the following it is convenient to introduce the notation

$$S_\lambda(v_0, \ldots, v_n) = \text{conv}\left\{\frac{1-\lambda}{n+1}\sum_{i=0}^{n} v_i + \lambda v_0, \ldots, \frac{1-\lambda}{n+1}\sum_{i=0}^{n} v_i + \lambda v_n\right\},$$

where λ is a rational number.

Theorem 3. *Let $K \subset \mathbb{R}^n$ be a convex body and $S := \text{conv}\{v_0, \ldots, v_n\}$ a simplex of minimal volume containing K. Then*

$$S_{-1/n}(v_0, \ldots, v_n) \subseteq K \subseteq S. \tag{8}$$

The vertices of $S_{-1/n}(v_0, \ldots, v_n)$ are the centroids of the facets of S.

Proof. The proof relies on a theorem by V. Klee [12] which states that the centroids $c_j = (\sum_{i \neq j} v_i)/n$ of the facets of the minimal simplex S are contained in K. Hence the theorem follows from

$$\frac{1+\frac{1}{n}}{n+1}\sum_{i=0}^{n} v_i - \frac{1}{n}v_j = c_j. \qquad \square$$

The theorem is illustrated by Figure 3. It is easily seen that (8) is not sufficient for minimality. In fact the same property holds if $S_{-1/n}(v_0, \ldots, v_n)$ is the simplex of maximal volume contained in K (see [13], Theorem 3(a)).

Using the shallow-cut simplices method, it is possible to compute a simplex in polynomial time which almost satisfies the property (8). It is given as follows:

Shallow-cut simplices method with rounding

Input: Affinely independent vectors $v_0^{[0]}, \ldots, v_n^{[0]} \in \mathbb{Q}^n$ such that

$$K \subseteq S^{[0]} := \operatorname{conv}\{v_0^{[0]}, \ldots, v_n^{[0]}\},$$

a rational positive number ε, a rational number β with $0 \leq \beta < 1/n$, and an integral positive number N such that

$$N > \frac{3(1+\beta)^2(n+1)^2}{(1-\beta n)^2} \ln\left(\frac{\operatorname{vol}(S^{[0]})}{\varepsilon}\right).$$

Step 1:   Set $k := 0$.

Step 2:   Determine the centroid of $S^{[k]}$

$$c^{[k]} := \frac{1}{n+1} \sum_{i=0}^{n} v_i^{[k]}.$$

Step 3a: Set $i := 0$.

Step 3b: Call the oracle for the separation problem with the input vector $(1+\beta)c^{[k]} - \beta v_i^{[k]}$. If the oracle finds a vector $a \in \mathbb{Q}^n$ such that $K \subset \{x \in \mathbb{R}^n \mid a^T x \leq a^T((1+\beta)c^{[k]} - \beta v_i^{[k]})\}$, then go to Step 4.

Step 3c: If $i = n$, then STOP $(S_{-\beta}(v_0^{[k]}, \ldots, v_n^{[k]}) \subseteq K)$, otherwise set $i := i+1$ and go to Step 3b.

Step 4:   Determine an index $l \in \{0, \ldots, n\}$ such that

$$a^T v_l^{[k]} = \min_{i \in \{0, \ldots, n\}} a^T v_i^{[k]}.$$

Step 5:   Determine a new simplex $S^{[k+1]} := \operatorname{conv}\{v_0^{[k+1]}, \ldots, v_n^{[k+1]}\}$ given by

$$v_i^{[k+1]} := (1 - \bar{\delta}_i) v_l^{[k]} + \bar{\delta}_i v_i^{[k]},$$

where

$$\bar{\delta}_i := 2^{-t} \lceil \delta_i 2^t \rceil,$$

$$\delta_i := \left(1 - \frac{1-\beta n}{n(n-\beta)} \frac{a^T(c^{[k]} - v_i^{[k]}) + \beta a^T(c^{[k]} - v_l^{[k]})}{(1+\beta)a^T(c^{[k]} - v_l^{[k]})}\right)^{-1},$$

$$t := 3\langle n \rangle + \left\lceil \log_2\left(\frac{1+\beta n}{(1-\beta n)^2}\right)\right\rceil.$$

Step 6:   If $k = N$, then STOP $(\operatorname{vol}(K) < \varepsilon)$, otherwise set $k := k+1$ and go to Step 2.

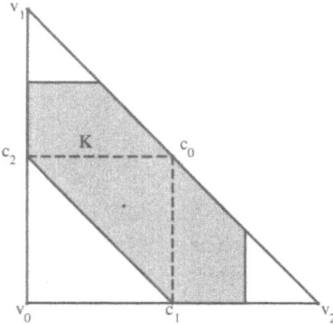

FIGURE 3. A minimal simplex containing K

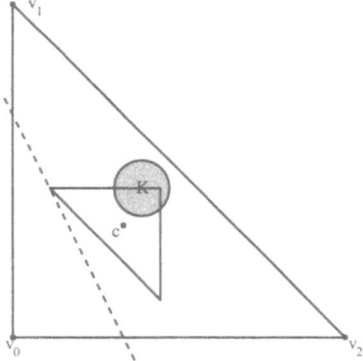

FIGURE 4. A shallow cut for $\beta = 1/3$

Note that by the convexity of K the method stops in Step 3c if and only if

$$S_{-\beta}(v_0^{[k]}, \ldots, v_n^{[k]}) \subseteq K. \tag{9}$$

If the algorithm proceeds to Step 4, then

$$K \subset \{x \in \mathbb{R}^n \mid a^T x \le a^T((1+\beta)c^{[k]} - \beta v_l^{[k]})\}.$$

This is illustrated by Figure 4. For $\beta = 0$ the algorithm coincides with simplices method with rounding as given in Section 4.

Theorem 4. *The shallow-cut simplices method with rounding finds a simplex*

$$S^{[k]} := \mathrm{conv}\{v_0^{[k]}, \ldots, v_n^{[k]}\}$$

such that either

(i) $S_{-\beta}(v_0^{[k]}, \ldots, v_n^{[k]}) \subseteq K \subseteq S^{[k]}$ *(the algorithm stops in Step 3c), or*
(ii) $K \subseteq S^{[k]}$ *and* $\mathrm{vol}(S^{[k]}) < \varepsilon$ *(the algorithm stops in Step 6).*

If the components of $v_0^{[0]}, \dots, v_n^{[0]}$ are integral, then the overall storage requirement is bounded from above by

$$n(n+1)\left(N(3\langle n\rangle + 2\langle\lceil 1/(1-\beta n)\rceil\rangle) + \max_{j\in\{0,\dots,n\}}\langle\|v_j^{[0]}\|_\infty\rangle\right)$$

bits.

The proof of the Theorem 4 is divided into several lemmas. Most of the proofs follow their counterparts for the original simplices method in [6] and [22]. By neglecting the rounding procedure ($t = 0$) or considering the central-cut version ($\beta = 0$), most of the proofs simplify by a large amount. Throughout it is assumed that $n \geq 2$.

Lemma 7. *In the shallow-cut simplices method with rounding, the relation*

$$S^{[k]} \cap \{x \in \mathbb{R}^n \mid a^T x \leq (1+\beta)a^T c^{[k]} - \beta a^T v_l^{[k]}\} \subseteq S^{[k+1]}$$

holds.

Proof. Let $x \in S^{[k]}$, i.e. $x = \sum_{i=0}^{n} \lambda_i v_i^{[k]}$ with $\sum_{i=0}^{n} \lambda_i = 1$ and $\lambda_i \geq 0$ for all $i \in \{0, \dots, n\}$. Setting

$$\mu_i := \frac{\lambda_i}{\bar{\delta}_i} \quad \text{for } i \in \{0, \dots, n\}\setminus\{l\}, \quad \mu_l := 1 - \sum_{i\neq l}\frac{\lambda_i}{\bar{\delta}_i},$$

it is easily checked that

$$\sum_{i=0}^{n} \mu_i v_i^{[k+1]} = x, \quad \sum_{i=0}^{n} \mu_i = 1.$$

Since $\bar{\delta}_i \geq \delta_i > 0$, it follows that $\mu_i \geq 0$ for all $i \in \{0, \dots, n\}\setminus\{l\}$. Finally we have

$$\mu_l = 1 - \sum_{i\neq l}\frac{\lambda_i}{\bar{\delta}_i} \geq \lambda_l + \frac{1-\beta n}{n(n-\beta)}\frac{\sum_{i\neq l}\lambda_i(a^T(c^{[k]} - v_i^{[k]}) + \beta a^T(c^{[k]} - v_l^{[k]}))}{(1+\beta)a^T(c^{[k]} - v_l^{[k]})}.$$

Hence, using $a^T x = \sum_{i=0}^{n} \lambda_i a^T v_i^{[k]}$, the inequality $a^T x \leq (1+\beta)a^T c^{[k]} - \beta a^T v_l^{[k]}$ implies that

$$\mu_l \geq \lambda_l + \frac{1-\beta n}{n(n-\beta)}\frac{(1-\lambda_l)((1+\beta)a^T c^{[k]} - \beta a^T v_l^{[k]}) - a^T x + \lambda_l a^T v_l^{[k]}}{(1+\beta)a^T(c^{[k]} - v_l^{[k]})}$$

$$\geq \lambda_l\left(1 - \frac{1-\beta n}{n(n-\beta)}\right) \geq 0. \qquad \square$$

Lemma 8. *Let $n \geq 2$ and $\beta \in [0, 1/n)$. Then*

$$\left(\frac{n^2-1}{n(n-\beta)}\right)^{n-1}\frac{n+1}{n(1+\beta)} > \varepsilon^{\frac{(1-\beta n)^2}{2(1+\beta n)(n+1)^2}}.$$

Proof. The left-hand-side equals

$$\left(1 - \frac{1 - \beta n}{n(n - \beta)}\right)^{n-1} \left(1 + \frac{1 - \beta n}{n(1 + \beta)}\right).$$

Taking logarithms, we obtain

$$s_n := (n - 1) \ln\left(1 - \frac{1 - \beta n}{n(n - \beta)}\right) + \ln\left(1 + \frac{1 - \beta n}{n(1 + \beta)}\right).$$

Expanding the logarithms into power series, this yields

$$
\begin{aligned}
s_n &= -\sum_{k=1}^{\infty} \frac{(n - 1)(1 - \beta n)^k}{n^k(n - \beta)^k k} - \sum_{k=1}^{\infty} \frac{(-1)^k(1 - \beta n)^k}{n^k(1 + \beta)^k k} \\
&= \sum_{k=1}^{\infty} \frac{(1 - \beta n)^k}{n^k k}\left(-\frac{n - 1}{(n - \beta)^k} - \frac{(-1)^k}{(1 + \beta)^k}\right) \\
&= \sum_{k \text{ odd}} \frac{(1 - \beta n)^k}{n^k}\left(\frac{1}{(1 + \beta)^k}\left(\frac{1}{k} - \frac{1 - \beta n}{(k + 1)n(1 + \beta)}\right)\right. \\
&\qquad \left. - \frac{n - 1}{(n - \beta)^k}\left(\frac{1}{k} + \frac{1 - \beta n}{(k + 1)n(n - \beta)}\right)\right).
\end{aligned}
\tag{10}
$$

The summand in (10) for $k = 1$ evaluates to

$$
\begin{aligned}
&\frac{(1 - \beta n)^2((1 + 2\beta)(n^2 - n + 1) - 3\beta^2 n)}{2n^2(1 + \beta)^2(n - \beta)^2} \\
\geq\ & \frac{(1 - \beta n)^2(1 + 2\beta)(n - 1)}{2n(1 + \beta)^2(n - \beta)^2} \\
=\ & \frac{(1 - \beta n)^2(1 + 2\beta)(n - 1)}{2(1 + \beta)^2((n - 1)(n + 1 - \beta)^2 - n^2 + n + (1 - \beta)^2} \\
\geq\ & \frac{(1 - \beta n)^2(1 + 2\beta)}{2(1 + \beta)^2(n + 1 - \beta)^2} \geq \frac{(1 - \beta n)^2}{2(1 + \beta)^2(n + 1)^2}.
\end{aligned}
$$

Using elementary calculus, it can be shown that for $k \geq 3$ and nonnegative β the term

$$\frac{1}{(1 + \beta)^k}\left(\frac{1}{k} - \frac{1 - \beta n}{(k + 1)n(1 + \beta)}\right)$$

decreases monotonously and for $\beta < \frac{1}{n}$ the term

$$\frac{n - 1}{(n - \beta)^k}\left(\frac{1}{k} + \frac{1 - \beta n}{(k + 1)n(n - \beta)}\right)$$

increases monotonously in β. Hence for any $k \in \mathbb{N}$ the summand in (10) is greater or equal than

$$\frac{(1 - \beta n)^k}{n^k k}\left(\left(\frac{n}{n+1}\right)^k - (n - 1)\left(\frac{n}{n^2 - 1}\right)^k\right) = \frac{(1 - \beta n)^k}{(n+1)^k k}\left(1 - \frac{1}{(n - 1)^{k-1}}\right) \geq 0.$$

The result follows. $\square$

Lemma 9. *Let $\beta \in [0, 1/n)$. Then in the shallow cut simplices method with rounding, the volumes of the simplices satisfy*

$$\frac{\text{vol}(S^{[k+1]})}{\text{vol}(S^{[k]})} \leq (1+\alpha)^n e^{-\frac{(1-\beta n)^2}{2(1+\beta)^2(n+1)^2}},$$

where $\alpha := (1 + 1/n)2^{-t}$.

Proof. Using that $a^T v_i^{[k]} \geq a^T v_l^{[k]}$ and

$$a^T v_i^{[k]} = (n+1)a^T c^{[k]} - \sum_{j \neq i} a^T v_j^{[k]} \leq (n+1)a^T c^{[k]} - n\, a^T v_l^{[k]},$$

it follows that

$$\frac{n(1+\beta)}{n+1} \leq \delta_i \leq \frac{n(n-\beta)}{n^2-1}.$$

Similarly,

$$\begin{aligned}
\sum_{i \neq l} \frac{1}{\delta_i} &= n - \frac{1 - \beta n}{n(n-\beta)} \frac{\sum_{i \neq l}(a^T(c^{[k]} - v_i^{[k]}) + \beta a^T(c^{[k]} - v_l^{[k]}))}{(1+\beta)a^T(c^{[k]} - v_l^{[k]})} \\
&= n - \frac{1 - \beta n}{n(n-\beta)} \frac{n((1+\beta)a^T c^{[k]} - \beta a^T v_l^{[k]}) - (n+1)a^T c^{[k]} + a^T v_l^{[k]}}{(1+\beta)a^T(c^{[k]} - v_l^{[k]})} \\
&= n + \frac{(1 - \beta n)^2}{n(n-\beta)(1+\beta)}.
\end{aligned}$$

Hence

$$\begin{aligned}
\frac{1}{\delta_i} &> \frac{1}{\delta_i + 2^{-t}} \geq \frac{1}{n(n-\beta)/(n^2-1) + 2^{-t}} \\
&= \frac{n^2-1}{n(n-\beta) + n(n-1)\alpha} > \frac{n^2-1}{n(n-\beta)(1+\alpha)} \qquad (11)
\end{aligned}$$

and

$$\begin{aligned}
\sum_{i \neq l} \frac{1}{\delta_i} &> \sum_{i \neq l} \frac{1}{\delta_i + 2^{-t}} = \sum_{i \neq l} \frac{1}{\delta_i(1 + 2^{-t}/\delta_i)} \\
&\geq \frac{1}{1 + 2^{-t}(n+1)/(n(1+\beta))} \left(\sum_{i \neq l} \frac{1}{\delta_i}\right) \\
&\geq \frac{1}{1+\alpha} \left(n + \frac{(1 - \beta n)^2}{n(n-\beta)(1+\beta)}\right). \qquad (12)
\end{aligned}$$

Each vector $v_i^{[k+1]}$ lies on the halfline from $v_l^{[k]}$ through $v_i^{[k]}$, and

$$\frac{\|v_i^{[k+1]} - v_l^{[k]}\|_2}{\|v_i^{[k]} - v_l^{[k]}\|_2} = \bar{\delta}_i. \quad \text{Hence} \quad \frac{\text{vol}(S^{[k+1]})}{\text{vol}(S^{[k]})} = \prod_{i \neq l} \bar{\delta}_i.$$

Due to (11) and (12), this value can be bounded from above by the inverse of the optimal value of

$$\min \prod_{i \neq l} \tau_i$$

$$\text{s.t. } \tau_i \geq \frac{n^2 - 1}{(1+\alpha)n(n-\beta)} \quad \text{for all} \quad i \in \{0, \ldots, n\} \setminus \{l\}$$

$$\sum_{i \neq l} \tau_i \geq \frac{1}{1+\alpha} \left(n + \frac{(1-\beta n)^2}{n(n-\beta)(1+\beta)} \right).$$

There is an optimal solution with $n-1$ of the τ_i at their lower bounds. We therefore obtain

$$\frac{\text{vol}(S^{[k]})}{\text{vol}(S^{[k+1]})}$$
$$\geq \left(\frac{n^2 - 1}{(1+\alpha)n(n-\beta)} \right)^{n-1} \frac{1}{1+\alpha} \left(n + \frac{(1-\beta n)^2}{n(n-\beta)(1+\beta)} - \frac{(n-1)(n^2-1)}{n(n-\beta)} \right)$$
$$= \frac{1}{(1+\alpha)^n} \left(\frac{n^2 - 1}{n(n-\beta)} \right)^{n-1} \frac{n+1}{n(1+\beta)}.$$

Note that the inequality holds even if the last restriction of the optimization problem is redundant. The result follows by Lemma 8. $\qquad \square$

Lemma 10. *If the simplices method with rounding does not find a feasible point $c^{[k]}$ within* $\left\lceil \dfrac{3(1+\beta)^2(n+1)^2}{(1-\beta n)^2} \ln \left(\dfrac{\text{vol}(S^{[0]})}{\varepsilon} \right) \right\rceil$ *iterations, then* $\text{vol}(K) < \varepsilon$.

Proof. By Lemma 9, the volume of $S^{[k]}$ can be bounded from above by

$$\text{vol}(S^{[k]}) \leq \text{vol}(S^{[0]})(1+\alpha)^{kn} e^{-k \frac{(1-\beta n)^2}{2(1+\beta)^2(n+1)^2}}.$$

This implies that $\text{vol}(S^{[k]}) < \varepsilon$ if the condition

$$k \left(\frac{(1-\beta n)^2}{2(1+\beta)^2(n+1)^2} - n \ln(1+\alpha) \right) > \ln \left(\frac{\text{vol}(S^{[0]})}{\varepsilon} \right)$$

is satisfied. Using that $\ln(1+x) \leq x$ for all $x > -1$, we have

$$\frac{(1-\beta n)^2}{2(1+\beta)^2(n+1)^2} - n \ln(1+\alpha) \geq \frac{(1-\beta n)^2}{2(1+\beta)^2(n+1)^2} - (n+1)2^{-t}$$

$$\geq \frac{(1-\beta n)^2}{2(1+\beta)^2(n+1)^2} - \frac{(1-\beta n)^2}{8(1+\beta)^2(n+1)^2}$$

$$> \frac{(1-\beta n)^2}{3(1+\beta)^2(n+1)^2}.$$

Since by Lemma 7 $K \subseteq S^{[k]}$, the result follows. $\qquad \square$

Lemma 11. *For all* $i \in \{0, \ldots, n\}$ *the vectors* $v_i^{[k]}$ *satisfy*

$$\|v_i^{[k]}\|_\infty \le 2^k \max_{j \in \{0,\ldots,n\}} \|v_j^{[0]}\|_\infty.$$

Proof. By definition we have

$$
\begin{aligned}
\|v_i^{[k+1]}\|_\infty &\le |1 - \bar{\delta}_i| \|v_l^{[k]}\|_\infty + \bar{\delta}_i \|v_i^{[k]}\|_\infty \\
&\le \max\{2\bar{\delta}_i - 1, 1\} \max_{j \in \{0,\ldots,n\}} \|v_j^{[[k]}\|_\infty \\
&< (2\left(\frac{n^2}{n^2 - 1} + 2^{-t}\right) - 1) \max_{j \in \{0,\ldots n\}} \|v_j^{[k]}\|_\infty \\
&\le \left(\frac{n^2 + 1}{n^2 - 1} + \frac{(1 - \beta n)^2}{4(1 + \beta)^2(n + 1)^3}\right) \max_{j \in \{0,\ldots,n\}} \|v_j^{[k]}\|_\infty \\
&< 2 \max_{j \in \{0,\ldots,n\}} \|v_j^{[k]}\|_\infty.
\end{aligned}
$$

The result follows.   □

With these lemmas, we can now prove the main theorem of the section.

Proof of Theorem 4. Since

$$K \subseteq S^{[k]} \cap \{x \in \mathbb{R}^n \mid a^T x \le (1 + \beta)a^T c^{[k]} - \beta a^T v_l^{[k]}\},$$

the Lemmas 7 and 10 show that the algorithm delivers a correct result. We now estimate the number of digits of $v_i^{[k]}$. Using Lemma 11, the number of digits before the decimal point of each component is bounded from above by

$$\log_2\left(2^k \max_{j \in \{0,\ldots,n\}} \|v_j^{[0]}\|_\infty + 1\right) < k + \max_{j \in \{0,\ldots,n\}} \langle \|v_j^{[0]}\|_\infty \rangle.$$

Using that the components of $v_0^{[0]}, \ldots, v_n^{[0]}$ are integral, the number of digits behind the decimal point is bounded from above by tk. Since

$$t < 3\langle n \rangle + 2\langle \lceil 1/(1 - \beta n) \rceil \rangle - 1$$

and since the vectors $v_0^{[k]}, \ldots, v_n^{[k]}$ have $n(n + 1)$ components altogether, the result follows.   □

6. Applications

In this section two applications of the shallow-cut simplices method are considered. In order to shorten the notation, we define

$$
\begin{aligned}
S_{-\beta} &:= S_{-\beta}(v_0, \ldots, v_n), \\
S &:= \text{conv}\{v_0, \ldots, v_n\}, \\
c &:= \frac{1}{n+1} \sum_{i=0}^{n} v_i.
\end{aligned}
$$

6.1. Volume computation

If any simplex S satisfying the property $S_{-\beta} \subseteq K \subseteq S$ is known, one immediately gets a lower and an upper bound on the volume of K. The relative error of these bounds (i.e. $\mathrm{vol}(S)/\mathrm{vol}(S_{-\beta})$) is greater than or equal to n^n which might appear to be very bad. However, it can be shown that any polynomial algorithm for computing bounds for the volume from an oracle for the corresponding separation problem has a relative error of at least $(cn/\log(n))^n$, where c is some positive constant [2]. Using the shallow-cut simplices method, one immediately gets bounds with a relative error that is arbitrarily close to n^n. This compares favourably with the relative error $n^{3n/2}$ from a corresponding algorithm based on the ellipsoid method (see [9]). The method is, however, worse than an algorithm by U. Betke and M. Henk [4] that produces bounds with a relative error that is arbitrarily close to $n!$. A different approach based on Markov-chains was used in algorithms given in [7] and [14]. The bounds from these methods have a relative error that is arbitrarily small, but they can only be guaranteed with a certain probability.

Corollary 2. *For some polytope $P \subset \mathbb{R}^n$, assume that the assumptions of Corollary 1 are satisfied. Then if the shallow-cut simplices method is called with the input*

$$
\begin{aligned}
v_0^{[0]} &:= R[-1,\ldots,-1]^T, \\
v_i^{[0]} &:= v_0^{[0]} + 2nRe_i \quad \text{for} \quad i \in \{1,\ldots,n\}, \\
\varepsilon &:= (2/r)^n, \\
\beta &:= \frac{n^{s-1}}{n^s+1}, \\
N &:= 12n(1+n)^2 n^{2s}(\langle R\rangle + \langle r\rangle + \langle n\rangle),
\end{aligned}
$$

where $s \in \mathbb{N}$ is some fixed number, then the volume of P is bounded by

$$\beta^n \mathrm{vol}(S^{[k]}) \leq \mathrm{vol}(K) \leq \mathrm{vol}(S^{[k]}), \tag{13}$$

where

$$\mathrm{vol}(S^{[k]}) = \frac{\det([v_1^{[k]} - v_0^{[k]}, \ldots, v_n^{[k]} - v_0^{[k]}])}{n!}.$$

The overall complexity for determining these bounds is polynomial in n, $\langle R\rangle$, $\langle r\rangle$, ϕ, and the number of oracle calls.

Proof. Since by assumption $\mathrm{vol}(P) \geq \varepsilon$, the shallow-cut simplices method stops in Step 3c. Then the property $S_{-\beta} \subseteq P \subseteq S$ directly yields (13). The complexity follows from $1/(1-\beta n) = n^s+1$ and the fact that the determinant of a matrix can be computed polynomially in the digital size of the matrix (see, e.g., [9] Corollary 1.4.9). $\square$

The constant $s \in \mathbb{N}$ can be chosen such that the relative error of the bounds for the volume comes arbitrarily close to n^n.

6.2. Total approximation problems

A second application of the simplices method lies in determining bounds for the optimal function value of a concave minimization problem

$$\min f(x)$$
$$\text{s.t. } x \in K, \qquad\qquad\qquad \text{(CPP)}$$

where $f : \mathbb{R}^n \to \mathbb{R}$ is a concave function and $K \subset \mathbb{R}^n$ is a convex body. This problem has been studied extensively, and there exist many algorithms for obtaining solutions for specific problems of the type above (see, e.g., [10]). However, with the exception of some algorithms for very special concave programming problems, these algorithms are not polynomial.

Provided that an oracle for the separation problem for the set K is given, one immediately gets bounds for the optimal function value of (CPP). Assume that the shallow-cut simplices method stops with the vertices $v_0, \ldots, v_n$ and $S_{-\beta} \subseteq K \subseteq S$. It follows that the optimal function value $f(x^*)$ can be bounded by

$$\min_{x \in S} f(x) \leq f(x^*) \leq \min_{x \in S_{-\beta}} f(x). \qquad\qquad (15)$$

It is easily seen that the optimal solution of (CPP) is attained at an extreme point of K (see, e.g., [10] Theorem I.1), and thus (15) can be simplified to

$$\min_{i \in \{0,\ldots,n\}} f(v_i) \leq f(x^*) \leq \min_{i \in \{0,\ldots,n\}} f((1+\beta)c - \beta v_i). \qquad (16)$$

Hence rough bounds for $f(x^*)$ can be obtained with $2(n+1)$ additional function evaluations.

We now focus on a special concave programming problem, namely the total approximation problem

$$\min \quad \|Ax\|_R$$
$$\text{s.t.} \quad \|x\|_D = 1, \qquad\qquad\qquad \text{(TAP)}$$

where $A \in \mathbb{Z}^{m \times n}$ with $m \geq n$, and where $\|\cdot\|_R$ and $\|\cdot\|_D$ are some given norms on $\mathbb{R}^m$ and $\mathbb{R}^n$, respectively. The problem (TAP) arises from the problem of fitting a hyperplane to a given set of points (see [3], [15], [19], [20], [21]). If the matrix A has rank n, then the optimal vectors of (TAP) and of (CPP) with

$$f(x) := -\|x\|_D \quad \text{and} \quad K := \{x \in \mathbb{R}^n \mid \|Ax\|_R \leq 1\}$$

are multiples of each other. The optimal function value of (CPP) is the negative inverse of the optimal value of (TAP).

In most applications the norms $\|\cdot\|_R$ and $\|\cdot\|_D$ are chosen as ℓ_1-, ℓ_2- or ℓ_∞-norms. The complexity of (TAP) depends on the norm used. In particular, if $\|\cdot\|_D = \|\cdot\|_\infty$, then the problem can be solved either by solving n systems of linear equations (for $\|\cdot\|_R = \|\cdot\|_2$) or by solving n linear programs (for $\|\cdot\|_R = \|\cdot\|_1$ or $\|\cdot\|_R = \|\cdot\|_\infty$, see [3]). For these cases the problem (TAP) can thus be solved by an algorithm that is polynomial in $L := \sum_{i=1}^m \sum_{j=1}^n \langle A_{ij} \rangle$.

If $\| \cdot \|_D$ is any other ℓ_p-norm, there is no polynomial algorithm available. However, it is possible to estimate the relative error of the bounds that can be obtained from (17). We consider the case that both $\| \cdot \|_R$ and $\| \cdot \|_D$ are ℓ_p-norms with $p \in \mathbb{N} \cup \{\infty\}$. First of all it is easily seen that $\{x \in \mathbb{R}^n \mid \|Ax\|_R \le 1\}$ is bounded if and only if the matrix A has rank n. This can be checked in polynomial time (see [9] Corollary 1.4.9). If A is rank-deficient, the optimal value of (TAP) is zero, so assume from now on that $\mathrm{rank}(A) = n$.

In order to use the shallow-cut simplices method, we must find appropriate expressions for the input and an algorithm for the corresponding separation problem. The latter is given as follows:

Algorithm for the **separation problem** with
$K := \{x \in \mathbb{R}^n \mid \|Ax\|_p \le 1\}, A \in \mathbb{Z}^{m \times n}, p \in \mathbb{N} \cup \{\infty\}$.

Input: A vector $c \in \mathbb{Q}^n$.

Step 1: If $p < \infty$, then compute $\|Ac\|_p^p$. If $p = \infty$, then compute $\|Ac\|_\infty$. If the computed value is smaller or equal to 1, then STOP (assert that $c \in K$).

Step 2: If $p = 1$, then compute $w \in \mathbb{Q}^m$ by $w_i := \mathrm{sign}(Ac)_i$.
If $2 \le p < \infty$, then compute $w \in \mathbb{Q}^m$ by $w_i := (Ac)_i^{p-1}$.
If $p = \infty$, then compute $w \in \mathbb{Q}^m$ by $w_i := 1$ for
$(Ac)_i = \|Ac\|_\infty, w_i := -1$ for $(Ac)_i = -\|Ac\|_\infty, w_i := 0$ otherwise.

Step 3: Compute $a := A^T w$ and STOP (a defines a separating hyperplane).

The vector w is chosen as a positive multiple of a subgradient of $\| \cdot \|_p$ at the point Ac, i.e. $\lambda w \in \partial \|Ac\|_p$ for some $\lambda > 0$ (see, e.g., [18]). By definition we then have

$$\|Ax\|_p \ge (\lambda w)^T A(x - c) + \|Ac\|_p$$

for all $x \in \mathbb{R}^n$. If $\|Ax\|_p \le 1$, it follows that $(A^T w)^T x \le (A^T w)^T c$ as required.

Theorem 5. *Consider the total approximation problem (TAP) with $A \in \mathbb{Z}^{m \times n}$ and $L := \sum_{i=1}^m \sum_{j=1}^n \langle A_{ij} \rangle$. Then if for some constant $s \in \mathbb{N}$ the shallow-cut simplices method using the separation oracle above is called with the input*

$$
\begin{aligned}
v_0^{[0]} &:= 2^L[-1, \ldots, -1]^T, \\
v_i^{[0]} &:= v_0^{[0]} + 2^{L+1} n e_i \quad \text{for} \quad i \in \{1, \ldots, n\}, \\
\varepsilon &:= 2^{-n(L-1)}/n!, \\
\beta &:= \frac{n^{s-1}}{n^s + 1}, \\
N &:= \frac{9}{2}(n^s + n^{s-1} + 1)(n+1)^3 L,
\end{aligned}
$$

it stops in Step 3c. If x^* is an optimal solution for (TAP), then

$$\left(\max_{i\in\{0,\dots n\}} ||v_i||_D\right)^{-1} \leq ||Ax^*||_R \leq \left(\max_{i\in\{0,\dots,n\}} ||(1+\beta)c - \beta v_i||_D\right)^{-1}.$$

The relative error of these bounds does not exceed $\dfrac{(3-\beta)n}{\beta(n+1)}$. *The overall algorithm is polynomial in L.*

Since the algorithm can only stop in Step 3c, it is obviously sufficient to omit the test in Step 6 of the shallow-cut simplices method, and therefore the values for N and ε do not have to be computed. The proof of Theorem 5 is split into several lemmas.

Lemma 12. *Let $A \in \mathbb{Z}^{m\times n}$ and $L := \sum_{i=1}^{m}\sum_{j=1}^{n}\langle A_{ij}\rangle$. Then*

$$\{x \in \mathbb{R}^n \mid ||x||_1 \leq 2^{-L}\} \subseteq \{x \in \mathbb{R}^n \mid ||Ax||_p \leq 1\} \subseteq \{x \in \mathbb{R}^n \mid ||x||_\infty \leq 2^L\}$$

for any $p \geq 1$.

Proof. On the one hand, we have

$$\begin{aligned}
\{x \in \mathbb{R}^n \mid ||Ax||_p \leq 1\} &\supseteq \{x \in \mathbb{R}^n \mid ||Ax||_1 \leq 1\} \\
&\supseteq \{x \in \mathbb{R}^n \mid ||x||_1 \leq ||A||_1^{-1}\} \\
&\supseteq \{x \in \mathbb{R}^n \mid ||x||_1 \leq 2^{-L}\}.
\end{aligned}$$

On the other hand, we have $\{x \in \mathbb{R}^n \mid ||Ax||_p \leq 1\} \subseteq \{x \in \mathbb{R}^n \mid ||Ax||_\infty \leq 1\}$. At least one optimal vector x^* of

$$\begin{aligned}
\max \quad &||x||_\infty \\
\text{s.t.} \quad &||Ax||_\infty \leq 1
\end{aligned}$$

is attained at a vertex of $\{x \in \mathbb{R}^n \mid ||Ax||_\infty \leq 1\}$, i.e., there exist n linear independent rows of A such that $a_i^T x^* \in \{-1,1\}$. It follows by Cramer's Rule that the coordinates of x^* can be expressed by

$$x_i^* = \frac{\det(A_\mathcal{I} \leftarrow_i e)}{\det A_\mathcal{I}},$$

where $A_\mathcal{I} \in \mathbb{Z}^{n\times n}$ is the square matrix consisting of the rows above and where $e \in \{-1,1\}^n$ is defined by $e_i := a_i^T x^*$. Since $|\det(D)| \leq 2^{\sum_{i,j}\langle D_{ij}\rangle - n^2}$ for all $D \in \mathbb{Z}^{n\times n}$ (see [9] Lemma 1.3.3), it follows that

$$|\det(A_\mathcal{I} \leftarrow_i e)| \leq 2^{L-n^2+n} \leq 2^L.$$

By $|\det(A_\mathcal{I})| \geq 1$, the result follows. $\square$

The following lemma states a property of the output of the shallow-cut simplices method if the set K is centrally symmetric (i.e. if it follows from $x \in K$ that $-x \in K$).

Lemma 13. *Let $K \subset \mathbb{R}^n$ be a centrally symmetric convex body with*

$$S_{-\beta}(v_0, \ldots, v_n) \subseteq K \subseteq \mathrm{conv}\{v_0, \ldots, v_n\},$$

where $0 \leq \beta \leq 1/n$. Then $0 \in S_{(1-\beta)/2}(v_0, \ldots, v_n)$.

Proof. By the symmetry of K the relation $-S_{-\beta} \subseteq S$ holds which in turn is equivalent to

$$-\frac{1+\beta}{1-\beta} c \in S.$$

For, if on the one hand $-(1+\beta)/(1-\beta)c \in S$, then

$$-(1+\beta)c + \beta x = (1-\beta)\left(-\frac{1+\beta}{1-\beta}\right)c + \beta x \in S$$

for all $x \in S$. On the other hand, assume that $-S_{-\beta} \subseteq S$ holds and that $-(1+\beta)/(1-\beta)c \notin S$. Then

$$-\frac{1+\beta}{1-\beta}(1-\beta^k)c \notin S$$

for $k \in \mathbb{N}$ large enough. Therefore by $-S_{-\beta} \subseteq S$

$$\frac{1}{\beta}\left((1+\beta)c - \frac{1+\beta}{1-\beta}(1-\beta^k)c\right) = -\frac{1+\beta}{1-\beta}(1-\beta^{k-1})c \notin S.$$

Repeating this argument gives $0 \notin S$, which contradicts the fact that K is centrally symmetric and nonempty.

If $-(1+\beta)/(1-\beta)c \in S$, then

$$\frac{1+\beta}{2}c + \frac{1-\beta}{2}x = 0$$

for some $x \in S$. The result follows. $\qquad\square$

Lemma 14. *Assume that*

$$S_{-\beta}(v_0, \ldots, v_n) \subseteq \{x \in \mathbb{R}^n \mid \|Ax\|_R \leq 1\} \subseteq \mathrm{conv}\{v_0, \ldots, v_n\},$$

where $0 \leq \beta \leq 1/n$. Then if x^* is an optimal solution for (TAP), the inequalities

$$\left(\max_{i \in \{0,\ldots,n\}} \|v_i\|_D\right)^{-1} \leq \|Ax^*\|_R \leq \left(\max_{i \in \{0,\ldots,n\}} \|(1+\beta)c - \beta v_i\|_D\right)^{-1}$$

hold. The relative error of these bounds does not exceed $\dfrac{(3-\beta)n}{\beta(n+1)}$.

Proof. By the homogenity of norms, $x^*/\|Ax^*\|_R$ is a maximizer of

$$\max \quad \|x\|_D$$
$$\mathrm{s.t.} \quad \|Ax\|_R \leq 1.$$

Hence, using (16),

$$\max_{i \in \{0,\ldots,n\}} \|(1+\beta)c - \beta v_i\|_D \leq \frac{\|x^*\|_D}{\|Ax^*\|_R} \leq \max_{i \in \{0,\ldots,n\}} \|v_i\|_D.$$

The first result follows. The bound on the relative error remains to be shown. On the one hand, we have

$$\max_{i\in\{0,...,n\}} \|v_i\|_D \le \max_{i\in\{0,...,n\}} \|v_i - c\|_D + \|c\|_D$$

By Lemma 13

$$0 = c + \frac{1-\beta}{2}(x - c) \quad \text{for some} \quad x \in S,$$

and therefore

$$\|c\|_D \le \frac{1-\beta}{2} \max_{i\in\{0,...,n\}} \|v_i - c\|_D.$$

Hence

$$\max_{i\in\{0,...,n\}} \|v_i\|_D \le \frac{3-\beta}{2} \max_{i\in\{0,...,n\}} \|v_i - c\|_D. \tag{18}$$

On the other hand,

$$\max_{i\in\{0,...,n\}} \|(1+\beta)c - \beta v_i\|_D$$

$$= \max_{x\in S} \|(1+\beta)c - \beta x\|_D$$

$$\ge \min_{z\in\mathbb{R}^n} \max_{x\in S} \|z - \beta(x - c)\|_D$$

$$\ge \min_{z\in\mathbb{R}^n} \max_{i\in\{0,...,n\}} \max\{\|z - \beta(v_i - c)\|_D, \|z + \beta(v_i - c)/n\|_D\}$$

$$\ge \frac{\beta(n+1)}{2n} \max_{i\in\{0,...,n\}} \|v_i - c\|_D. \tag{19}$$

Combining (18) andd (19), the second result follows.   □

Proof of Theorem 5. The validity of the input vectors $v_0^{[0]}, \ldots, v_n^{[0]}$ easily follows by Lemma 12. Since $\text{vol}(P) \ge \varepsilon$, the method cannot stop in Step 6. The value of N simply follows by substitution. Since the number of digits of the vectors $(1+\beta)c^{[k]} - \beta v_i^{[k]}$ is polynomially bounded by t, N, and L, the given algorithm for the separation problem and hence the overall algorithm is polyomial in L. The bounds including the estimation for the relative error are given by Lemma 14. □

Acknowledgements. I acknowledge the support of the German-Israeli Foundation. I thank G.M. Ziegler for numerous valuable comments and suggestions on the manuscript.

References

[1] M. Akgül, *On Yamnitsky-Levin algorithm*, Cah. Cent. Etud. Rech. Oper. 26 (1984), 179–194.

[2] I. Barany and Z. Füredi, *Computing the volume is difficult*, Discrete Comput. Geom. 2 (1987), 319–326.

[3] S.G. Bartels, *Totale Approximationsprobleme*, Wissenschaft und Technik Verlag, Berlin 1995.

[4] U. Betke and M. Henk, *Approximating the volume of convex bodies*, Discrete Comput. Geom. 10 (1993), 15–21.

[5] R.G. Bland, D. Goldfarb and M.J. Todd, *The ellipsoid method: a survey*, Oper. Res. 29 (1981), 1039–1091.

[6] V. Chvátal, *Linear Programming*, Freeman Press, New York, 1983.

[7] M. Dyer, A. Frieze and R. Kannan, *A random polynomial-time algorithm for approximating the volume of convex bodies*, J. Assoc. Comput. Mach. 38 (1991), 1–17.

[8] U. Faigle, M. Hunting and W. Kern, *On a variant of the ellipsoid method: using simplices instead of ellipsoids*, extended abstract for the SOR conference 1995 in Passau.

[9] M. Grötschel, L. Lovász and A. Schrijver, *Geometric Algorithms and Combinatorial Optimization*, Springer-Verlag, Berlin 1988; 2nd edition, 1993.

[10] R. Horst and H. Tuy, *Global Optimization*, Springer-Verlag, Berlin, 1990.

[11] L.G. Khachian, *A polynomial algorithm in linear programming*, Sov. Math. Dokl. 20 (1979), 191–194.

[12] V. Klee, *Facet centroids and volume minimization*, Studia Sci. Math. Hungar. 21 (1986), 143–147.

[13] J.C. Lagarias and G.M. Ziegler, *Bounds for lattice polytopes containing a fixed number of interior points in a sublattice*, Can. J. Math. 43 (1991), 1022–1035.

[14] L. Lovász and M. Simonovits, *Random walks in a convex body and an improved volume algorithm*, Random Struct. Algorithms 4 (1993), 359–412.

[15] M.R. Osbourne and G.A. Watson, *An analysis of the total approximation problem in separable norms, and an algorithm for the total ℓ_1-problem*, SIAM J. Sci. Stat. Comp. 6 (1985), 410–425.

[16] M. Padberg, *Linear Optimization and Extensions*, Springer-Verlag, Berlin, 1995.

[17] R. Prakash and K.J. Supowit, *Increasing the rate of convergence of Yamnitsky and Levin's algorithm*, Working Paper Ohio State University, 1991.

[18] R.T. Rockafellar, *Convex Analysis*, Princeton University Press, 1970.

[19] H. Späth and G.A. Watson, *On orthogonal linear ℓ_1-approximation*, Numer. Math. 52 (1987), 531–543.

[20] S. Van Huffel and J. Vandewalle, *The Total Least Squares Problem*, SIAM Frontiers in Applied Mathematics 9, 1991.

[21] G.A. Watson, *The total approximation problem*, in: Approximation Theory IV (C.K. Chui, L.L. Schumaker, J.D. Ward, eds.), Academic Press, New York, 1983, 723–728.

[22] B. Yamnitsky and L.A. Levin, *An old linear programming algorithm runs in polynomial time*, in: 23rd Annual Symposium on Foundations of Computer Science, IEEE, New York, 1982, 327–328.

Sven G. Bartels
Emmerich-Josef-Str. 61
D-65929 Frankfurt/M.